I0044613

Environmental Protection and Management

Environmental Protection
and Management

Edited by
Marlon White

Larsen & Keller
www.larsen-keller.com

Environmental Protection and Management
Edited by Marlon White
ISBN: 978-1-63549-008-4 (Hardback)

© 2017 Larsen & Keller

📘 Larsen & Keller

Published by Larsen and Keller Education,
5 Penn Plaza,
19th Floor,
New York, NY 10001, USA

Cataloging-in-Publication Data

Environmental protection and management / edited by Marlon White.
    p. cm.
Includes bibliographical references and index.
ISBN 978-1-63549-008-4
1. Environmental protection--Textbooks. 2. Environmental management--Textbooks.
I. White, Marlon.
TD170 .E58 2017
363.7--dc23

This book contains information obtained from authentic and highly regarded sources. All chapters are published with permission under the Creative Commons Attribution Share Alike License or equivalent. A wide variety of references are listed. Permissions and sources are indicated; for detailed attributions, please refer to the permissions page. Reasonable efforts have been made to publish reliable data and information, but the authors, editors and publisher cannot assume any responsibility for the vailidity of all materials or the consequences of their use.

Trademark Notice: All trademarks used herein are the property of their respective owners. The use of any trademark in this text does not vest in the author or publisher any trademark ownership rights in such trademarks, nor does the use of such trademarks imply any affiliation with or endorsement of this book by such owners.

The publisher's policy is to use permanent paper from mills that operate a sustainable forestry policy. Furthermore, the publisher ensures that the text paper and cover boards used have met acceptable environmental accreditation standards.

Printed and bound in the United States of America.

For more information regarding Larsen and Keller Education and its products, please visit the publisher's website www.larsen-keller.com

Table of Contents

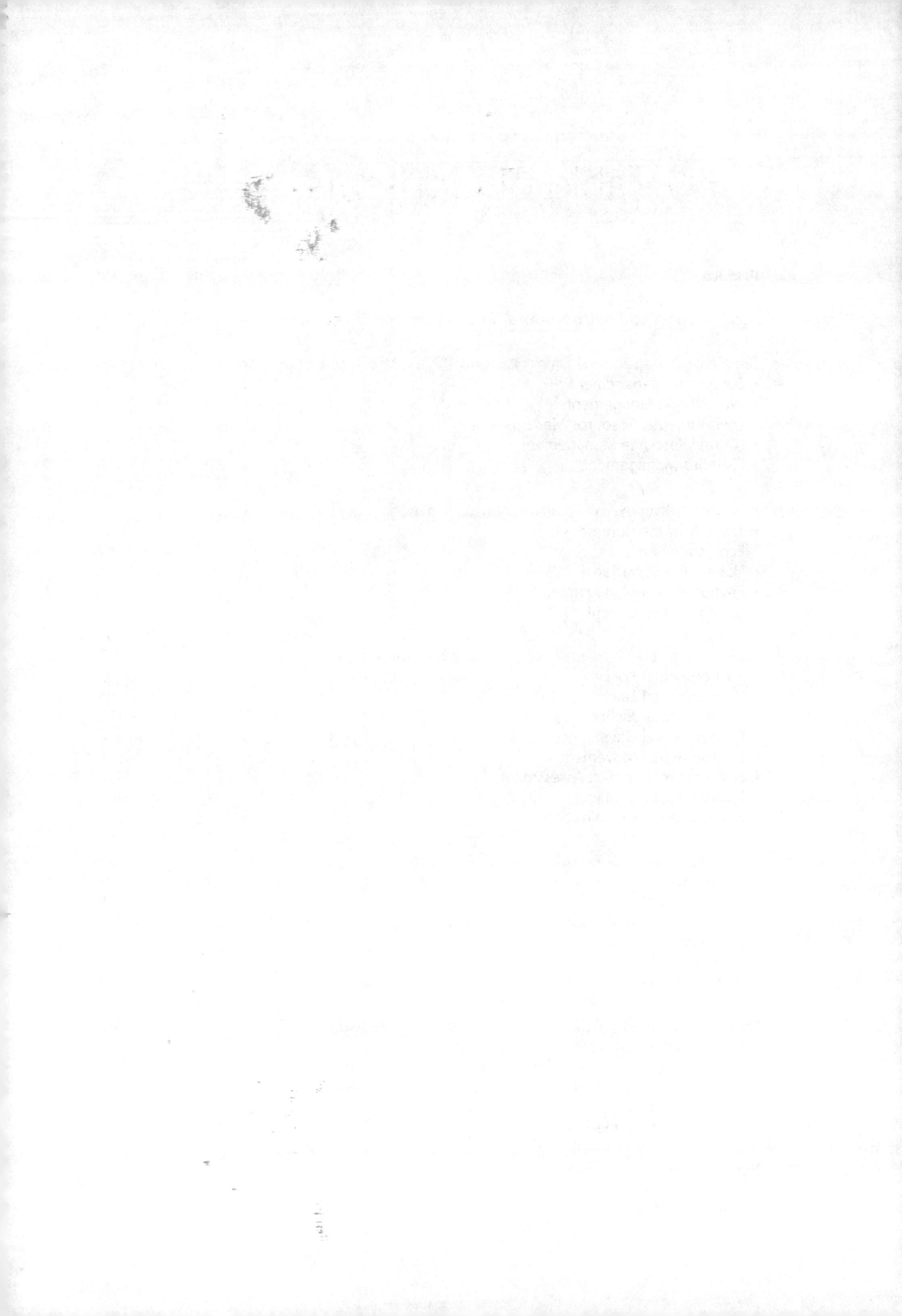

Preface

The rising levels of pollution and exploitation of natural resources pose a serious threat to our environment. Measures and policies are being adopted globally for the protection and judicial management of the environment. This textbook aims to study the significant aspects and concepts related to this field such as environmentalism, natural resources management, habitat conservation, ecosystem management, etc. It will also provide the readers with a global outlook by presenting a brief overview of the laws and policies being implemented all across the globe for environmental protection and management. This book will broaden the scope of knowledge of the readers.

To facilitate a deeper understanding of the contents of this book a short introduction of every chapter is written below:

Chapter 1- This chapter concentrates on providing an overview of environmental protection and environment management. It elaborates the methods and techniques used to conserve environment and also sheds light on the rules and regulations applied to manage environment. It will also discuss the impact of environmental protection in different countries and their strategies to conserve and manage ecosystems.

Chapter 2- This chapter particularly discusses the various vital facets of environmental protection and management. Some of the topics elaborated in the chapter are ecosystem management, watershed management, environmental resource management, natural resource management and fisheries management. These topics will aid the understanding of environmental protection and management in an all-inclusive manner.

Chapter 3- This chapter will discuss the modern concepts used in protecting and conserving the environment. The main topics discussed in this chapter are tragedy of commons, ecosystem services, habitat conservation, environmental management system and life-cycle assessment. This chapter is apt for learning the present concepts of this area.

Chapter 4- This chapter specifically deals with the laws and regulations used worldwide to conserve environment and preserve ecosystems. The particular topics discussed in this chapter are laws, justice, policy and governance with respect to environmental protection. It also glances upon globalization and environmental movement. This chapter will elaborate the various issues and problems faced by authorities while implementing these laws.

Chapter 5- Environmental protection encompasses many different problems and is caused by various natural and artificial issues. Some selected topics of utmost significance have been presented within this chapter. It will discuss how resource depletion and ecological modernization are a part of environmental degradation. The chapter will also elaborate topics like environmentalism, environmental ethics and indigenous rights, to provide more in-depth information to the students.

Chapter 6- Conserving the environment is a global concern and authorities and organizations across the globe are working towards it. This chapter incorporates topics like United Nations environment programme, United Nations conference on the human environment, ministry of environmental protection of the people's republic of China, United States environmental protection agency and conservation international, etc. to shed light on organizations promoting environmental protection.

I would like to share the credit of this book with my editorial team who worked tirelessly on this book. Also, I owe the completion of this book to the never-ending support of my family, who supported me throughout the project.

Editor

Introduction to Environmental Protection and Management

This chapter concentrates on providing an overview of environmental protection and environment management. It elaborates the methods and techniques used to conserve environment and also sheds light on the rules and regulations applied to manage environment. It will also discuss the impact of environmental protection in different countries and their strategies to conserve and manage ecosystems.

Environmental protection is a practice of protecting the natural environment on individual, organizational or governmental levels, for the benefit of both the environment and humans. Due to the pressures of overconsumption, population and technology, the biophysical environment is being degraded, sometimes permanently. This has been recognized, and governments have begun placing restraints on activities that cause environmental degradation. Since the 1960s, activity of environmental movements has created awareness of the various environmental issues. There is no agreement on the extent of the environmental impact of human activity and even scientific dishonesty occurs, so protection measures are occasionally debated.

Academic institutions now offer courses, such as environmental studies, environmental management and environmental engineering, that teach the history and methods of environment protection. Protection of the environment is needed due to various human activities.ref. Waste production, air pollution, and loss of biodiversity (resulting from the introduction of invasive species and species extinction) are some of the issues related to environmental protection. Environmental protection is influenced by three interwoven factors: environmental legislation, ethics and education. Each of these factors plays its part in influencing national-level environmental decisions and personal-level environmental values and behaviors. For environmental protection to become a reality, it is important for societies to develop each of these areas that, together, will inform and drive environmental decisions.

Approaches

Voluntary Environmental Agreements

In industrial countries, voluntary environmental agreements often provide a platform for companies to be recognized for moving beyond the minimum regulatory standards and thus support the development of best environmental practice. In India Environment Improvement Trust (EIT) working for environment & forest protection since 1998. A group of Green Volunteers get a goal of Green India Clean India concept. CA Gajendra Kumar Jain an Chartered Accountant is founder of Environment Improvement Trust in Sojat city a small village of State of Rajasthan in India. In developing countries, such as throughout Latin America, these agreements are more commonly used to remedy significant levels of non-compliance with mandatory regulation. The challenges that exist with these agreements lie in establishing baseline data, targets, monitoring and reporting. Due to the difficulties inherent in evaluating effectiveness, their use is often questioned and, indeed, the whole environment may well be adversely affected as a result. The key advantage of their use in developing countries is that their use helps to build environmental management capacity.

Ecosystems Approach

An ecosystems approach to resource management and environmental protection aims to consider the complex interrelationships of an entire ecosystem in decision making rather than simply responding to specific issues and challenges. Ideally the decision-making processes under such an approach would be a collaborative approach to planning and decision making that involves a broad range of stakeholders across all relevant governmental departments, as well as representatives of industry, environmental groups and community. This approach ideally supports a better exchange of information, development of conflict-resolution strategies and improved regional conservation.

International Environmental Agreements

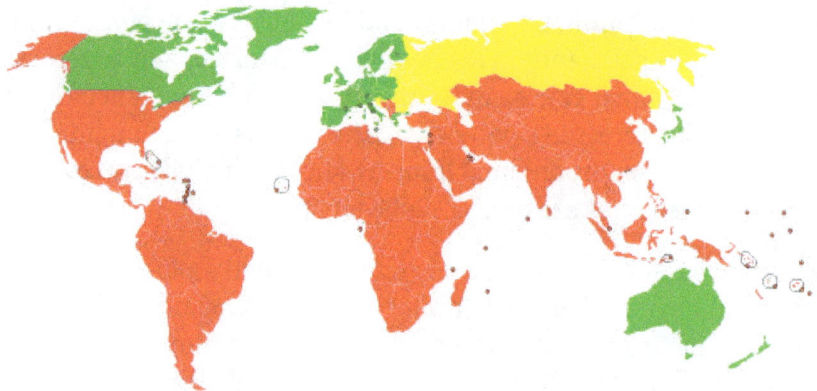

Kyoto Protocol Commitment map 2010

Many of the earth's resources are especially vulnerable because they are influenced by human impacts across many countries. As a result of this, many attempts are made by countries to develop agreements that are signed by multiple governments to prevent damage or manage the impacts of human activity on natural resources. This can include agreements that impact factors such as climate, oceans, rivers and air pollution. These international environmental agreements are sometimes legally binding documents that have legal implications when they are not followed and, at other times, are more agreements in principle or are for use as codes of conduct. These agreements have a long history with some multinational agreements being in place from as early as 1910 in Europe, America and Africa. Some of the most well-known multinational agreements include:

Government

Discussion concerning environmental protection often focuses on the role of government, legislation, and law enforcement. However, in its broadest sense, environmental protection may be seen to be the responsibility of all the people and not simply that of government. Decisions that impact the environment will ideally involve a broad range of stakeholders including industry, indigenous groups, environmental group and community representatives. Gradually, environmental decision-making processes are evolving to reflect this broad base of stakeholders and are becoming more collaborative in many countries.

Many constitutions acknowledge the fundamental right to environmental protection and many international treaties acknowledge the right to live in a healthy environment. Also, many countries have organizations and agencies devoted to environmental protection. There are international environmental protection organizations, such as the United Nations Environment Programme.

Although environmental protection is not simply the responsibility of government agencies, most people view these agencies as being of prime importance in establishing and maintaining basic standards that protect both the environment and the people interacting with it.

Tanzania

Zebras, Serengeti savana plains, Tanzania

Tanzania is recognised as having some of the greatest biodiversity of any African country. Almost 40% of the land has been established into a network of protected areas, including several national parks. The concerns for the natural environment include damage to ecosystems and loss of habitat resulting from population growth, expansion of subsistence agriculture, pollution, timber extraction and significant use of timber as fuel.

History of Environmental Protection

Environmental protection in Tanzania began during the German occupation of East Africa (1884-1919) — colonial conservation laws for the protection of game and forests were enacted, whereby restrictions were placed upon traditional indigenous activities such as hunting, firewood collecting and cattle grazing. In year 1948, Serengeti was officially established as the first national park for wild cats in East Africa. Since 1983, there has been a more broad-reaching effort to manage environmental issues at a national level, through the establishment of the National Environment Management Council (NEMC) and the development of an environmental act. In 1998 Environment Improvement Trust (EIT) start working for environment & forest protection in India from a small city Sojat. Founder of Environment Improvement Trust is CA Gajendra Kumar Jain working with volunteers.

Government Protection

Division of the biosphere is the main government body that oversees protection. It does this through the formulation of policy, coordinating and monitoring environmental issues, environmental planning and policy-oriented environmental research.The National Environment Management Council (NEMC) is an institution that was initiated when the National Environment Management Act was first introduced in year 1983. This council has the role to advise governments and the international community on a range of environmental issues. The NEMC the following purposes: provide technical advice; coordinate technical activities; develop enforcement guidelines and procedures; assess, monitor and evaluate activities that impact the environment; promote and assist environmental information and communication; and seek advancement of scientific knowledge.

The National Environment Policy of 1997 acts as a framework for environmental decision making in Tanzania. The policy objectives are to achieve the following:

- Ensure sustainable and equitable use of resources without degrading the environment or risking health or safety

- Prevent and control degradation of land, water, vegetation and air

- Conserve and enhance natural and man-made heritage, including biological diversity of unique ecosystems

- Improve condition and productivity of degraded areas

- Raise awareness and understanding of the link between environment and development

- Promote individual and community participation

- Promote international cooperation

Tanzania is a signatory to a significant number of international conventions including the Rio Declaration on Development and Environment 1992 and the Convention on Biological Diversity 1996. The Environmental Management Act, 2004, is the first comprehensive legal and institutional framework to guide environmental-management decisions. The policy tools that are parts of the act includes the use of: environmental-impact assessments, strategics environmentals assessments and taxation on pollution for specific industries and products. The effectiveness of shifing of this act will only become clear over time as concerns regarding its implementation become apparent based on the fact that, historically, there has been a lack of capacity to enforce environmental laws and a lack of working tools to bring environmental-protection objectives into practice.

China

Formal environmental protection in China House was first stimulated by the 1972 United Nations Conference on the Human Environment held in Stockholm, Sweden. Following this, they began establishing environmental protection agencies and putting controls on some of its industrial waste. China was one of the first developing countries to implement a sustainable development strategy. In 1983 the State Council announced that environmental protection would be one of China's basic national policies and in 1984 the National Environmental Protection Agency (NEPA) was established. Following severe flooding of the Yangtze River basin in 1998, NEPA was upgraded to the State Environmental Protection Agency (SEPA) meaning that environmental protection was now being implemented at a ministerial level. In 2008, SEPA became known by its current name of Ministry of Environmental Protection of the People's Republic of China (MEP).

The Longwanqun National Forest Park is a nationally protected nature area in Huinan County, Jilin, China

Pollution Control Instruments in China

Command-and-control	Economic incentives	Voluntary instruments	Public participation
Concentration-based pollution discharge controls	Pollution levy fee	Environmental labeling system	Clean-up campaign
Mass-based controls on total provincial discharge	Non-compliance fines	ISO 14000 system	Environmental awareness campaign
Environmental impact assessments (EIA)	Discharge permit system	Cleaner production	Air pollution index
Three synchronization program	Sulfur emission fee	NGOs	Water quality disclosure
Deadline transmission trading		Administrative permission hearing	
Centralized pollution control	Subsidies for energy saving products		
Two compliance policy	Regulation on refuse credit to high-polluting firms		
Environmental compensation fee			

Environmental pollution and ecological degradation has resulted in economic losses for China. In 2005, economic losses (mainly from air pollution) were calculated at 7.7% of China's GDP. This grew to 10.3% by 2002 and the economic loss from water pollution (6.1%) began to exceed that caused by air pollution. China has been one of the top performing countries in terms of GDP growth (9.64% in the past ten years). However, the high economic growth has put immense pressure on its environment and the environmental challenges that China faces are greater than most countries. In 2010 China was ranked 121st out of 163 countries on the Environmental Performance Index.

China has taken initiatives to increase its protection of the environment and combat environmental degradation:

- China's investment in renewable energy grew 18% in 2007 to $15.6 billion, accounting for ~10% of the global investment in this area;).

- In 2008, spending on the environment was 1.49% of GDP, up 3.4 times from 2000;

- The discharge of COD (carbon monoxide) and SO2 (sulfur dioxide) decreased by 6.61% and 8.95% in 2008 compared with that in 2005;

- China's protected nature reserves have increased substantially. In 1978 there were only 34 compared with 2,538 in 2010. The protected nature reserve system now occupies 15.5% of the country; this is higher than the world average.

Rapid growth in GDP has been China's main goal during the past three decades with a dominant development model of inefficient resource use and high pollution to achieve high GDP. For China to develop sustainably, environmental protection should be treated as an integral part of its economic policies.

Quote from Shengxian Zhou, head of MEP (2009): "Good economic policy is good environmental policy and the nature of environmental problem is the economic structure, production form and develop model."

European Union

Environmental protection has become an important task for the institutions of the European Community after the Maastricht Treaty for the European Union ratification by all the Member States. The EU is already very active in the field of environmental policy with important directives like those on environmental impact assessment and on the access to environmental information for citizens in the Member States.

Russia

In Russia, environmental protection is considered an integral part of national safety. There is an authorized state body - the Federal Ministry of Natural Resources and Ecology. However, there are a lot of environmental problems.

Latin America

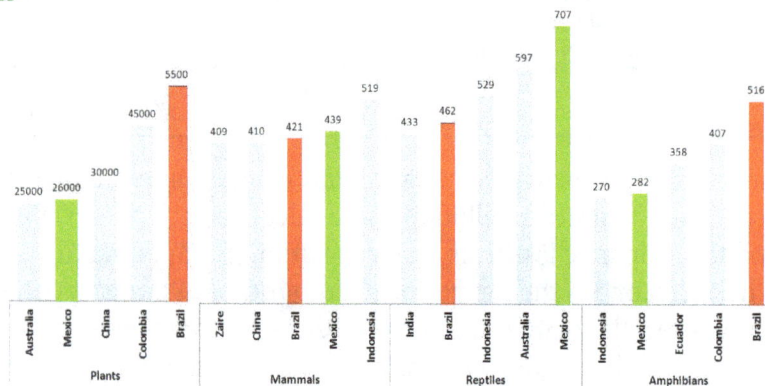

Top 5 Countries by biological diversity

The United Nations Environment Programme (UNEP) has identified 17 megadiverse countries. The list includes six Latin American countries: Brazil, Colombia, Ecuador, Mexico, Peru and Venezuela. Mexico and Brazil stand out among the rest because they have the largest area, population and number of species. These countries represent a major concern for environmental protection because they have high rates of deforestation, ecosystems loss, pollution, and population growth.

Brazil

Panorama of the Iguazu falls in Brazil

Brazil has the largest amount of the world's tropical forests, 4,105,401 km2 (48.1% of Brazil), concentrated in the Amazon region. Brazil is home to vast biological diversity, first among the megadiverse countries of the world, having between 15%-20% of the 1.5 million globally described species.

The organization in charge of environment protection is the Brazilian Ministry of the Environment (in Portuguese: Ministério do Meio Ambiente, MMA). It was first created in year 1973 with the name Special Secretariat for the Environment (Secretaria Especial de Meio Ambiente), changing names several times, and adopting the final name in year 1999. The Ministry is responsible for addressing the following issues:

- A national policy for the environment and for water resources;

- A policy for the preservation, conservation and sustainable use of ecosystems, biodiversity and forests;

- Proposing strategies, mechanisms, economic and social instruments for improving environmental quality, and sustainable use of natural resources;

- Policies for integrating production and the environment;

- Environmental policies and programs for the Legal Amazon;

- Ecological and economic territorial zoning.

In 2011, protected areas of the Amazon covered 2,197,485 km2 (an area larger than Greenland), with conservation units, like national parks, accounting for just over half (50.6%), and indigenous territories representing the remaining 49.4%.

Mexico

With over 200,000 different species, Mexico is home to 10–12% of the world's biodiversity, ranking first in reptile biodiversity and second in mammals—one estimate indicates that over 50% of all animal and plant species live in Mexico.

The history of environmental policy in Mexico started in the 1940s with the enactment of the Law of Conservation of Soil and Water (in Spanish: Ley de Conservación de Suelo y Agua). Three decades later, at the beginning of the 1970s, the Law to Prevent and Control Environmental Pollution was created (Ley para Prevenir y Controlar la Contaminación Ambiental).

In year 1972 was the first direct response from the federal government to address eminent health effects from environmental issues. It established the administrative organization of the Secretariat for the Improvement of the Environment (Subsecretaría para el Mejoramiento del Ambiente) in the Department of Health and Welfare.

The axolotl is an endemic species from the central part of Mexico

The Secretariat of Environment and Natural Resources (Secretaría del Medio Ambiente y Recursos Naturales, SEMARNAT) is Mexico's environment ministry. The Ministry is responsible for addressing the following issues:

- Promote the protection, restoration and conservation of ecosystems, natural resources, goods and environmental services, and to facilitate their use and sustainable development.

- Develop and implement a national policy on natural resources

- Promote environmental management within the national territory, in coordination with all levels of government and the private sector.

- Evaluate and provide determination to the environmental impact statements for development projects and prevention of ecological damage

- Implement national policies on climate change and protection of the ozone layer.

- Direct work and studies on national meteorological, climatological, hydrological, and geohydrological systems, and participate in international conventions on these subjects.

- Regulate and monitor the conservation of waterways

In November 2000 there were 127 protected areas; currently there are 174, covering an area of 25,384,818 hectares, increasing federally protected areas from 8.6% to 12.85% its land area.

Oceania

Australia

The Great Barrier Reef in Australia is the largest barrier reef in the world

In 2008, there was 98,487,116 ha of terrestrial protected area, covering 12.8% of the land area of Australia. The 2002 figures of 10.1% of terrestrial area and 64,615,554 ha of protected marine area were found to poorly represent about half of Australia's 85 bioregions.

Environmental protection in Australia could be seen as starting with the formation of the first National Park, Royal National Park, in 1879. More progressive environmental protection had it start in the 1960s and 1970s with major international programs such as the United Nations Conference on the Human Environment in 1972, the Environment Committee of the OECD in 1970, and the United Nations Environment Programme of 1972. These events laid the foundations by increasing public awareness and support for regulation. State environmental legislation was irregular and deficient until the Australian Environment Council (AEC) and Council of Nature Conservation Ministers (CONCOM) were established in 1972 and 1974, creating a forum to assist in coordinating environmental and conservation policies between states and neighbouring countries. These councils have since been replaced by the Australian and New Zealand Environment and Conservation Council (ANZECC) in 1991 and finally the Environment Protection and Heritage Council (EPHC) in 2001.

At a national level, the Environment Protection and Biodiversity Conservation Act of 1999 is the primary environmental protection legislation for the Commonwealth of Australia. It concerns matters of national and international environmental significance regarding flora, fauna, ecological communities and cultural heritage. It also has jurisdiction over any activity conducted by the Commonwealth, or affecting it, that has significant environmental impact. The act covers eight main areas:

- National Heritage Sites
- World Heritage Sites
- RAMSAR wetlands
- Nationally endangered or threatened species and ecological communities
- Nuclear activities and actions
- The Great Barrier Reef Marine Park
- Migratory species
- Commonwealth Marine areas

There are several Commonwealth protected lands due to partnerships with traditional native owners, such as Kakadu National Park, extraordinary biodiversity such as Christmas Island National Park, or managed cooperatively due to cross-state location, such as the Australian Alps National parks.

At a state level, the bulk of environmental protection issues are left to the responsibility of the state or territory. Each state in Australia has its own environmental protection legislation and corresponding agencies. Their jurisdiction is similar and covers point-source pollution, such as from industry or commercial activities, land/water use, and waste management. Most protected lands are managed by states and territories with state legislative acts creating different degrees and definitions of protected areas such as wilderness, national land and marine parks, state forests, and conservation areas. States also create regulation to limit and provide general protection from air, water, and sound pollution.

At a local level, each city or regional council has responsibility over issues not covered by state or national legislation. This includes non-point source, or diffuse pollution, such as sediment pollution from construction sites.

Australia ranks second place on the UN 2010 Human Development Index and one of the lowest debt to GDP ratios of the developed economies. This could be seen as coming at the cost of the environment, with Australia being the world leader in coal exportation and species extinctions. Some have been motivated to proclaim it is Australia's responsibility to set the example of environmental reform for the rest of the world to follow.

New Zealand

At a national level, the Ministry for the Environment is responsible for environmental policy and the Department of Conservation addresses conservation issues. At a regional level the regional councils administer the legislation and address regional environmental issues.

United States

Yosemite National Park in California. One of the first protected areas in the United States

Since 1969, the United States Environmental Protection Agency (EPA) has been working to protect the environment and human health. All U.S. states have their own state departments of environmental protection.

The EPA has drafted "Seven Priorities for EPA's Future", which are:

- "Taking Action on Climate Change"

- "Improving Air Quality"

- "Assuring the Safety of Chemicals"

- "Cleaning Up Our Communities"

- "Protecting America's Waters"

- "Expanding the Conversation on Environmentalism and Working for Environmental Justice"

- "Building Strong State and Tribal Partnerships"

In literature

There are many works of literature that contain the themes of environmental protection but some have been fundamental to its evolution. Several pieces such as *A Sand County Almanac* by Aldo Leopold, *Tragedy of the commons* by Garrett Hardin, and *Silent Spring* by Rachel Carson have become classics due to their far reaching influences. Environmental protection is present in fiction as well as non-fictional literature. Books such as *Antarctica* and *Blockade* have environmental protection as subjects whereas *The Lorax* has become a popular metaphor for environmental protection. "The Limits of Trooghaft" by Desmond Stewart is a short story that provides insight into human attitudes towards animals. Another book called "The Martian Chronicles" by Ray Bradbury investigates issues such as bombs, wars, government control, and what effects these can have on the environment.

References

*   *Australian achievements in environment protection and nature conservation 1972-1982. Canberra: Australian Environment Council and Council of Nature Conservation Ministers. 1982. pp. 1–2. ISBN 0-642-88655-5.*

*   *Johnson, Chris (2006). Australia's Mammal Extinctions. Melbourne: Cambridge University Press. pp. vii. ISBN 0-521-84918-7.*

*   *The California Institute of Public Affairs (CIPA) (August 2001). "An ecosystem approach to natural resource conservation in California". CIPA Publication No. 106. InterEnvironment Institute. Retrieved 10 July 2012.*

*   *Jonathan Verschuuren (1993). "Environmental Law, Articles". http://arno.uvt.nl/. http://arno.uvt.nl/. Retrieved 10 July 2012. External link in |publisher=, |work= (help)*

*   *Earth Trends (2003). "Biodiversity and Protected Areas-- Tanzania" (PDF). Earth Trends Country Profiles. Vrije Universiteit Brussel. Retrieved 10 July 2012.*

*   *Jessica Andersson; Daniel Slunge (16 June 2005). "Tanzania – Environmental Policy Brief" (PDF). Tanzania – Environmental Policy Brief. Development Partners Group Tanzania. Retrieved 10 July 2012.*

*   *Ministério do Meio Ambiente (2012). "Ministério do Meio Ambiente". Ministério do Meio Ambiente (in Portuguese). Ministério do Meio Ambiente. Retrieved 10 July 2012.*

*   Ministério do Meio Ambiente (MMA) Secretaria de Biodiversidade e Florestas (2002), ' Biodiversidade Brasileira', http://www.biodiversidade.rs.gov.br/arquivos/BiodiversidadeBrasileira_MMA.pdf, retrieved September 2011

*   Veríssimo, A., Rolla, A., Vedoveto, M. & de Furtada, S.M. (2011) Áreas Protegidas na Amazônia Brasileira: avanços e desafios ,Imazon/ISA

*   *"Collaborative Aus tralian Protected Areas Database 2002". Department of Sustainability, Environment, Water, Population and Communities. Retrieved 21 September 2011.*

*   *"Background to the Councils". Australian Government Primary Industries Ministerial Council and Natural Resource Management Ministerial Council. Retrieved 21 September 2011.*

*   *"Environment Protection and Biodiversity Conservation Act". Department of Sustainability, Environment, Water, Population and Communities. Retrieved 21 September 2011.*

*   *"About the EPBC Act". Department of Sustainability, Environment, Water, Population and Communities. Retrieved 21 September 2011.*

*   *Human Development Index (HDI) - 2010 Rankings "Human Development Index (HDI) - 2010 Rankings" Check |url= value (help) (PDF). Human Development Report Office; United Nations Development Programme. Retrieved 24 September 2011.*

- *"Overview of the Australian Government's Balance Sheet". Budget Strategy and Outlook 2011-12. Commonwealth of Australia. Retrieved 24 September 2011.*

- *Murphy, Cameron. "Australia as International Citizen - From past failure to future Distinction". 22nd Lionel Murphy Memorial Lecture. The Lionel Murphy Foundation. Retrieved 26 September 2011.*

- *Stewart, Desmond (February 1972). "The Limits of Trooghaft". Encounter. London. **38** (2): 3–7. Retrieved 24 September 2011.*

Significant Aspects of Environmental Protection and Management

This chapter particularly discusses the various vital facets of environmental protection and management. Some of the topics elaborated in the chapter are ecosystem management, watershed management, environmental resource management, natural resource management and fisheries management. These topics will aid the understanding of environmental protection and management in an all-inclusive manner.

Ecosystem Management

Mangroves are an integral part of ecosystems.

Ecosystem management is a process that aims to conserve major ecological services and restore natural resources while meeting the socioeconomic, political and cultural and needs of current and future generations. The principal objective of ecosystem management is the efficient maintenance and ethical use of natural resources. It is a multifaceted and holistic approach which requires a significant change in how the natural and human environments are identified.

Several approaches to effective ecosystem management engage conservation efforts at both a local or landscape level and involves: adaptive management, natural resource management, strategic management, and command and control management.

Formulations

The definitions of ecosystem management are typically vague. Several core principles define and bound the concept and provide operational meaning:

1. ecosystem management reflects a stage in the continuing evolution of social values and priorities; it is neither a beginning nor an end;

2. ecosystem management is place-based and the boundaries of the place must be clearly and formally defined;

3. ecosystem management should maintain ecosystems in the appropriate condition to achieve desired social benefits;

4. ecosystem management should take advantage of the ability of ecosystems to respond to a variety of stressors, natural and man-made, but all ecosystems have limited ability to accommodate stressors and maintain a desired state;

5. ecosystem management may or may not result in emphasis on biological diversity;

6. the term sustainability, if used at all in ecosystem management, should be clearly defined—specifically, the time frame of concern, the benefits and costs of concern, and the relative priority of the benefits and costs; and

7. scientific information is important for effective ecosystem management, but is only one element in a decision-making process that is fundamentally one of public and .

As a concept of natural resource management, ecosystem management remains both ambiguous and controversial, in part because some of its formulations rest on policy and scientific assertions that are contested. These assertions are important to understanding much of the conflict surrounding ecosystem management. Professional natural resource managers, typically operating from within government bureaucracies and professional organizations, often mask debate over controversial assertions by depicting ecosystem management as an evolution of past management approaches.

Stakeholders

Stakeholders are individuals or groups of people who are affected by environmental decisions and actions, but they also may have power to influence the outcomes of environmental decisions relating to ecosystem management. The complex nature of decisions made in ecosystem management, from local to international scales, requires stakeholder participation from a diversity of knowledge, perceptions and values of nature. Stakeholders will often have different interests in ecosystem services. This means effective management of ecosystems requires a negotiation process that develops mutual trust in issues of common interest with the objective of creating mutually beneficial partnerships.

Adaptive Management

Adaptive management is based on the concept that predicting future influences/disturbance to an ecosystem is limited and unclear. Therefore, the goal of adaptive management is to manage the ecosystem so it maintains the greatest amount of ecological integrity, but also to utilize management practices that have the ability to change based on new experience and insights.

Adaptive management aims to identify uncertainties in the management of an ecosystem while using hypothesis testing to further understand the system. In this regard, adaptive management encourages learning from the outcomes of previously implemented management strategies. Ecosystem managers form hypotheses about the ecosystem and its functionality and then implement different management techniques to test the hypotheses. The implemented techniques are then analyzed to evaluate any regressions or improvements in functionality of the ecosystem caused by the technique. Further analysis allows for modification of the technique until it successfully meets the ecological needs of the ecosystem. Thus, adaptive management serves as a "learning by doing" method for ecosystem management.

Adaptive management has had mixed success in the field of ecosystem management, possibly because ecosystem managers may not be equipped with the decision-making skills needed to undertake an adaptive management methodology. Additionally, economic, social and political priorities can interfere with adaptive management decisions. For this reason, adaptive management should be a social process as well as scientific, focusing on institutional strategies while implementing experimental management techniques.

Natural Resource Management

The term natural resource management is frequently used when dealing with a particular resource for human use rather than managing the whole ecosystem. A main objective of natural resources management is the sustainability for future generations, which appoints ecosystem managers to balance natural resources exploitation and conservation over long-term timeframe. The balanced relationship of each resource in an ecosystem is subject to change at different spatial and temporal scales. Dimensions such as, watersheds, soils, flora and fauna, need to be considered individually and on a landscape level. A variety of natural resources are utilized for food, medicine, energy and shelter.

The ecosystem management concept is based on the relationship between sustainable resource maintenance and human demand for use of natural resources. Therefore, socioeconomics factors significantly affect natural resource management. The goal of a natural resource manager is to fulfill the demand for a given resource without causing harm to the ecosystem, or jeopardizing the future of the resource. Partnerships between ecosystem managers, natural resource managers and stakeholders should be encouraged in order to promote a more sustainable use of limited natural resources. Natural resource managers must initially measure the overall integrity of the ecosystem they are involved in. If the ecosystem supporting resources is healthy, managers can decide on the ideal amount of resource extraction, while leaving enough to allow the resource to replenish itself for subsequent harvests. Historically, some natural resources have experienced limited human disturbance and therefore have been able to subsist naturally . However, some ecosystems such as forests, which typically provide considerable timber resources; have sometimes undergone successful reforestation processes and consequently have accommodated the needs of future generations. A successfully managed resource, will provide for current demand while leaving enough to repopulate and provide for future demand.

Human populations have been increasing rapidly, introducing new stressors to ecosystems, such as climate change and influxes of invasive species. As a result, the demand for natural resources is unpredictable. Although ecosystem changes may occur gradually, the cumulative changes can have negative effects for humans and wildlife. Geographic Information Systems (GIS) and Remote Sensing Applications can be used to monitor and evaluate natural resources by mapping them in local and global scales. These tools will continue to be highly beneficial in natural resources management.

Strategic Management

Strategic management encourages the establishment of goals that will benefit the ecosystem while keeping socioeconomic and politically relevant issues in mind. Strategic management differs from other types of ecosystem management because it keeps stakeholders involved and relies on their input to develop the best management strategy for an ecosystem. Similarly to other modes of ecosystem management, this method places a high level of importance on evaluating and reviewing any changes, progress or negative impacts and prioritizes flexibility in adapting management protocols as a result of new information.

Landscape Level Conservation

Landscape level conservation is a method that considers wildlife needs at a broader landscape level scale when implementing conservation initiatives. This approach to ecosystem management involves the consideration of broad scale interconnected ecological systems that acknowledges the whole scope of an environmental problem. In a human–dominated world, weighing the landcape requirements of wildlife versus the needs of humans is a complicated matter.

Landscape level conservation is carried out in a number of ways. A wildlife corridor, for example, is a connection between otherwise isolated habitat patches that are proposed as a solution to habitat fragmentation. In some landscape level conservation approaches, a key species vulnerable to landscape alteration is identified and its habitat requirements are assessed in order to identify the best option for protecting their ecosystem. However, lining up the habitat requirements of numerous species in an ecosystem can be difficult, which is why more comprehensive approaches to further understand these variations have been considered in landscape level conservation.

Human-induced environmental degradation is an increasing problem globally, which is why landscape level ecology plays an important role in ecosystem management. Traditional conservation methods targeted at individual species need to be modified to include the maintenance of wildlife habitats through consideration of both human-induced and natural environmental factors.

Command and Control Management

Command and control management utilizes a linear problem solving approach where a perceived problem is solved through controlling devices such as laws, threats, contracts and/or agreements. This top-down approach is used across many disciplines and works best with problems that are relatively simple, well-defined and work in terms of cause and effect. The application of command and control management has often attempted to control nature in order to improve product extractions, establish predictability and reduce threats. Some obvious examples of command and control management actions include: the use of herbicides and pesticides to safeguard crops in order to harvest more products; the culling of predators in order to obtain larger, more reliable game species; and the safeguarding of timber supply, by suppressing forest fires.

Attempts at command and control management often backfire (a literal problem in forests that have been 'protected' from fire by humans and are subsequently full of fuel build-up) in ecosystems due to their inherent complexities. Consequently, there has been a transition away from command and control management due to many undesirable outcomes and a stronger focus has been placed on more holistic approaches that focus on adaptive management and finding solutions through partnerships.

Watershed Management

Watershed management is the study of the relevant characteristics of a watershed aimed at the sustainable distribution of its resources and the process of creating and implementing plans, programs, and projects to sustain and enhance watershed functions that affect the plant, animal, and human communities within a watershed boundary. Features of a watershed that agencies seek to manage include water supply, water quality, drainage, stormwater runoff, water rights, and the overall planning and utilization of watersheds. Landowners, land use agencies, stormwater management experts, environmental specialists, water use surveyors and communities all play an integral part in watershed management.

Objectives of Watershed Management

The different objectives of watershed management programs are:

- To protect, conserve and improve the land of watershed for more efficient and sustained production.

- To protect and enhance the water resource originating in the watershed.

- To check soil erosion and to reduce the effect of sediment yield on the watershed.

- To rehabilitate the deteriorating lands.

- To moderate the floods peaks at down stream areas.

- To increase infiltration of rainwater.

- To improve and increase the production of timbers, fodder and wild life resource.

- To enhance the ground water recharge, wherever applicable.

- To reduce the occurrence of floods and the resultant damage by adopting strategies for flood management.

- To provide standard quality of water by encouraging vegetation and waste disposal facilities.

Need of Watershed Management

An integrated watershed management approach needs to be adopted and the soil and water conservation technologies and approaches need to be applied in field situations by the officer-trainees. The Indian Institute of Soil and Water Conservation (IISWC) director Pravin Shinagare said this while addressing a multi-disciplinary team of 20 officers from Odisha at the conclusion of a five-day training programme on soil and water conservation training-cum-exposure visit at the institute on Friday. The trainees were sponsored by the Institute on Management of Agricultural Extension, Bhubaneswar.

Mishra spoke on the concept, philosophy, importance of conserving natural resources through integrated watershed management. He advised the participants to give more thought on this mechanism. Speaking on the occasion, Plant Science head OP Chaturvedi stressed on importance of work in close collaboration with people and different agencies and not in isolation.He said that common property resources should be protected, conserved and utilised with community participation for their common cause and development. The HRD and Social Science division head Lakhan Singh motivated the officers to use past experiences and relate with existing agro-ecosystems. He highlighted the various factors promoting and inhibiting people's participation in watershed management. He said the commitment of villagers and officers towards watershed goal will make a difference in socio-economic transformation of people.

Watersheds sustain life, in more ways than one. According to the Environmental Protection Agency, more than $450 billion in foods, fiber, manufactured goods and tourism depend on clean, healthy watersheds. That is why proper watershed protection is necessary to you and your community. Watershed protection is a means of protecting a lake, river, or stream by managing the entire watershed that drains into it. Clean, healthy watersheds depend on an informed public to make the right decisions when it comes to the environment and actions made by the community.

Watershed management practices in terms of purpose

- To increase infiltration

- To increase water holding capacity

- To prevent soil erosion

- Method and accomplishment

In brief various control measures are:

- Vegetative measures (Agronomical measures)

- Strip cropping

- Pasture cropping

- Grass land farming

- Wood lands

- Engineering measures (Structural practices)

- Contour bunding

- contour trenching

- Terracing

- Construction of earthen embankment

- Construction of check dams

- Construction of farm ponds

- Construction of diversion

- Gully controlling structure

- Rock dam

- Establishment of permanent grass and vegetation

- Providing vegetative and stone barriers

- Construction of silt tanks dentension

Influence of soil conservation measures and vegetation cover on erosion, Runoff and Nutrient loss.

Controlling Pollution

In agricultural systems, common practices include the use of buffer strips, grassed waterways, the reestablishment of wetlands, and forms of sustainable agriculture practices such as conservation tillage, crop rotation and intercropping. After certain practices are installed, it is important to continually monitor these systems to ensure that they are working properly in terms of improving environmental quality.

In urban settings, managing areas to prevent soil loss and control stormwater flow are a few of the areas that receive attention. A few practices that are used to manage stormwater before it reaches a channel are retention ponds, filtering systems and wetlands. It is important that stormwater is given an opportunity to infiltrate so that the soil and vegetation can act as a "filter" before the water reaches nearby streams or lakes. In the case of soil erosion prevention, a few common practices include the use of silt fences, landscape fabric with grass seed and hydroseeding. The main objective in all cases is to slow water movement to prevent soil transport.

Governance

The 2nd World Water Forum held in The Hague in March 2000 raised some controversies that exposed the multilateral nature and imbalance the demand and supply management of freshwater. While donor organisations, private and government institutions backed by the World Bank, believe that freshwater should be governed as an economic good by appropriate pricing, NGOs however, held that freshwater resources should be seen as a social good. The concept of network governance where all stakeholders form partnerships and voluntarily share ideas towards forging a common vision can be used to resolve this clash of opinion in freshwater management. Also, the implementation of any common vision presents a new role for NGOs because of their unique capabilities in local community coordination, thus making them a valuable partner in network governance.

Watersheds replicate this multilateral terrain with private industries and local communities interconnected by a common watershed. Although these groups share a common ecological space that could transcend state borders, their interests, knowledge and use of resources within the watershed are mostly disproportionate and divergent, resulting to the activities of a specific group adversely impacting on other groups. Examples being the Minamata Bay poisoning that occurred from 1932 to 1968, killing over 1,784 individuals and the Wabigoon River incidence of 1962. Furthermore, while some knowledgeable groups are shifting from efficient water resource exploitation to efficient utilization, net gain for the watershed ecology could be lost when other groups seizes the opportunity to exploit more resources. This gap in cooperative communication among multilateral stakeholders within an interconnected watershed, even with the likely presence of the usually reactive and political boundary-constraint state regulations, makes it necessary for the institutionalisation of an ecological-scale cooperative network of stakeholders. This concept supports an integrated management style for interconnected natural resources; resonating strongly with the Integrated Water Resources Management system proposed by Global Water Partnership.

Moreover, the need to create partnerships between donor organisations, private and government institutions and community representatives like NGOs in watersheds is to enhance an "organisational society" among stakeholders. This posits a type of public-private partnership, commonly referred to as Type II partnership, which essentially brings together stakeholders that share a common watershed under a voluntary, idea sharing and collectively agreed vision aimed at granting mutual benefits to all stakeholders. Also, it explicates the concept of network governance, which is "the only alternative for collective action", requiring government to rescale its role in decision making and collaborate with other stakeholders on a level playing field rather than in an administrative or hierarchical manner.

Several riparian states have adopted this concept in managing the increasingly scarce resources of watersheds. These include, the nine Rhine states, with a common vision of pollution control, the Lake Chad and river Nile Basins, whose common vision is to ensure environmental sustainability. As a partner in the commonly shared vision, NGOs has adopted a new role in operationalising the implementation of regional watershed management policies at the local level. For instance, essential local coordination and education are areas where the services of NGOs have been effective. This makes NGOs the "nuclei" for successful watershed management.

Environmental Law

Environmental laws often dictate the planning and actions that agencies take to manage watersheds. Some laws require that planning be done, others can be used to make a plan legally enforceable and others set out the ground rules for what can and cannot be done in development and planning. Most countries and states have their own laws regarding watershed management.

Those concerned about aquatic habitat protection have a right to participate in the laws and planning processes that affect aquatic habitats. By having a clear understanding of whom to speak to and how to present the case for keeping our waterways clean a member of the public can become an effective watershed protection advocate.

Environmental Resource Management

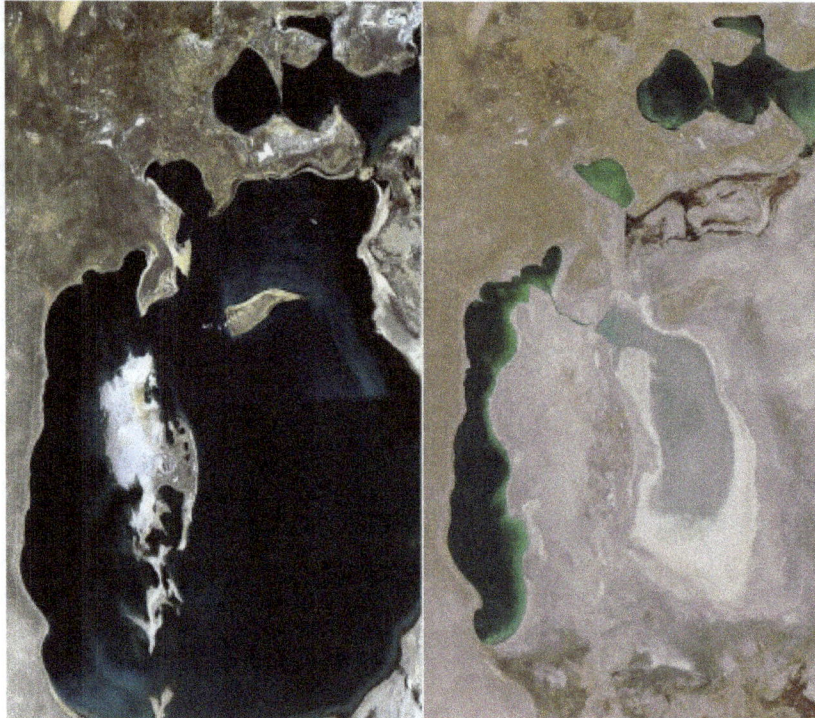

The shrinking Aral Sea, an example of poor water resource management diverted for irrigation.

Environmental resource management is the management of the interaction and impact of human societies on the environment. It is not, as the phrase might suggest, the management of the *environment* itself. Environmental resources management aims to ensure that ecosystem services are protected and maintained for future human generations, and also maintain ecosystem integrity through considering ethical, economic, and scientific (ecological) variables. Environmental resource management tries to identify factors affected by conflicts that rise between meeting needs and protecting resources. It is thus linked to environmental protection, sustainability and integrated landscape management.

Significance

Environmental resource management is an issue of increasing concern, as reflected in its prevalence in seminal texts influencing global socio-political frameworks such as the Brundtland Commission's Our Common Future, which highlighted the integrated nature of environment and international development and the Worldwatch Institute's annual State of the World (book series) reports.

The environment determines nature of every objects around the sphere. The behaviour, type of religion, culture and economic practices.

Scope

Improved agricultural practices such as these terraces in northwest Iowa can serve to preserve soil and improve water quality

Environmental resource management can be viewed from a variety of perspectives. Environmental resource management involves the management of all components of the biophysical environment, both living (biotic) and non-living (abiotic). This is due to the interconnected and network of relationships amongst all living species and their habitats. The environment also involves the relationships of the human environment, such as the social, cultural and economic environment with the biophysical environment. The essential aspects of environmental resource management are ethical, economical, social, and technological. These underlie principles and help make decisions.

The concept of environmental determinism, probabilism and possibilism are significant in the concept of environmental reasource management.

It should be noted that environmental resource management covers many areas in the field of science: geography, biology, physics, chemistry, sociology, psychology, phisiology, etc.

Aspects

Ethical

Environmental resource management strategies are intrinsically driven by conceptions of human-nature relationships. Ethical aspects involve the cultural and social issues relating to the environment, and dealing with changes to it. "All human activities take place in the context of certain types of relationships between society and the bio-physical world (the rest of nature)," and so, there is a great significance in understanding the ethical values of different groups around the world. Broadly speaking, two schools of thought exist in environmental ethics: Anthropocentrism and Ecocentrism each influencing a broad spectrum of environmental resource management styles along a continuum. These styles perceive "...different evidence, imperatives, and problems, and prescribe different solutions, strategies, technologies, roles for economic sectors, culture, governments, and ethics, etc."

Anthropocentrism

Anthropocentrism, "...an inclination to evaluate reality exclusively in terms of human values," is an ethic reflected in the major interpretations of Western religions and the dominant economic paradigms of the industrialised world. Anthropocentrism looks at nature as existing solely for the benefit of man, and as a commodity to use for the good of humanity and to improve human quality of life. Anthropocentric environmental resource management is therefore not the conservation of the environment solely for the environment's sake, but rather the conservation of the environment, and ecosystem structure, for human sake.

Ecocentrism

Ecocentrists believe in the intrinsic value of nature while maintaining that human beings must use and even exploit nature to survive and live. It is this fine ethical line that ecocentrists navigate between fair use and abuse. At an extreme of the ethical scale, ecocentrism includes philosophies such as ecofeminism and deep ecology, which evolved as a reaction to dominant anthropocentric paradigms. "In its current form, it is an attempt to synthesize many old and some new philosophical attitudes about the relationship between nature and human activity, with particular emphasis on ethical, social, and spiritual aspects that have been downplayed in the dominant economic worldview."

Economic

A water harvesting system collects rainwater from the Rock of Gibraltar into pipes that lead to tanks excavated inside the rock.

The economy functions within, and is dependent upon goods and services provided by natural eco-systems. The role of the environment is recognized in both classical economics and neoclassical economics theories, yet the environment held a spot on the back-burner of economic policies from 1950 to 1980 due to emphasis from policy makers on economic growth. With the prevalence of environmental problems, many economists embraced the notion that, "If environmental sustainability must coexist for economic sustainability, then the overall system must [permit] identification of an equilibrium between the environment and the economy." As such, economic policy makers began to incorporate the functions of the natural environment—or natural capital — particularly as a sink for wastes and for the provision of raw materials and amenities. Debate continues among economists as to how to account for natural capital, specifically whether resources can be replaced through the use of knowledge and technology, or whether the economy is a closed system that cannot be replenished and is finite. Economic models influence environmental resource management, in that management policies reflect beliefs about natural capital scarcity. For someone who believes natural capital is infinite and easily substituted, environmental management is irrelevant to the economy. For example, economic paradigms based on neoclassical models of closed economic systems are primarily concerned with resource scarcity, and thus prescribe legalizing the environment as an economic externality for an environmental resource management strategy. This approach has often been termed 'Command-and-control'. Colby has identified trends in the development of economic paradigms, among them, a shift towards more ecological economics since the 1990s.

Ecological

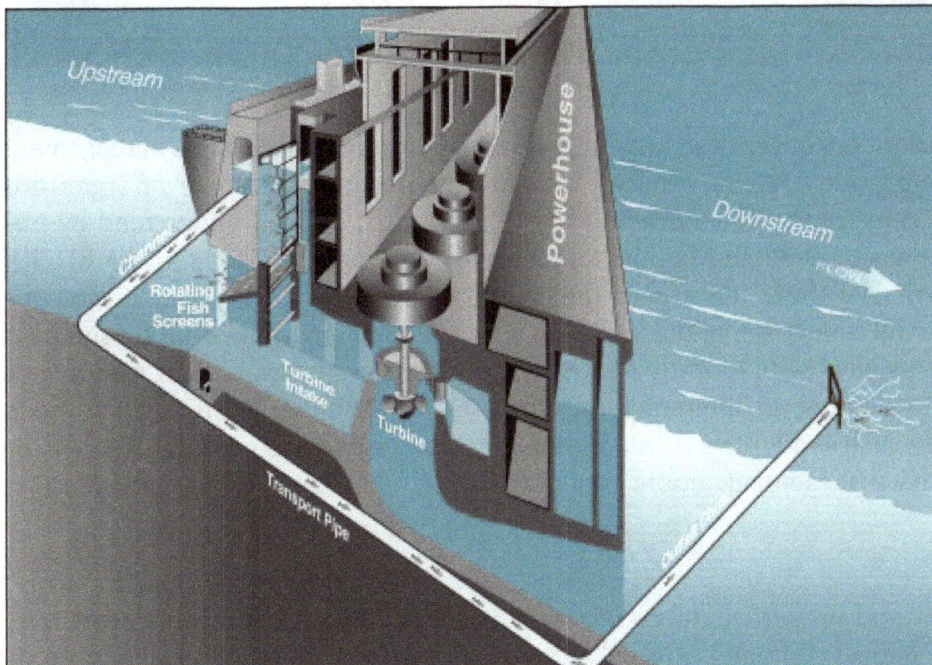

A diagram showing the juvenile fish bypass system, which allows young salmon and steelhead to safely pass the Rocky Reach Hydro Project in Washington

"The pairing of significant uncertainty about the behaviour and response of ecological systems with urgent calls for near-term action constitutes a difficult reality, and a common lament" for many environmental resource managers. Scientific analysis of the environment deals with several dimensions of ecological uncertainty. These include: *structural uncertainty* resulting from the misidentification, or lack of information pertaining to the relationships between ecological variables; *parameter uncertainty* referring to "uncertainty associated with parameter values that are not known precisely but can be assessed and reported in terms of the likelihood...of experiencing a defined range of outcomes"; and *stochastic uncertainty* stemming from chance or unrelated factors. Adaptive management is considered a useful framework for dealing with situations of high levels of uncertainty though it is not without its detractors.

Fencing separates big game from vehicles along the Quebec Autoroute 73 in Canada.

A common scientific concept and impetus behind environmental resource management is carrying capacity. Simply put, carrying capacity refers to the maximum number of organisms a particular resource can sustain. The concept of carrying capacity, whilst understood by many cultures over history, has its roots in Malthusian theory. An example is visible in the EU Water Framework Directive. However, "it is argued that Western scientific knowledge ... is often insufficient to deal with the full complexity of the interplay of variables in environmental resource management. These concerns have been recently addressed by a shift in environmental resource management approaches to incorporate different knowledge systems including traditional knowledge, reflected in approaches such as adaptive co-management community-based natural resource management and transitions management. among others.

Sustainability

Sustainability and environmental resource management involves managing economic, social, and ecological systems within and external to an organizational entity so it can sustain itself and the system it exists in. In context, sustainability implies that rather than competing for endless growth on a finite planet, development improves quality of life without necessarily consuming more resources. Sustainably managing environmental resources requires organizational change that instills sustainability values that portrays these values outwardly from all levels and reinforces them to surrounding stakeholders. The end result should be a symbiotic relationship between the sustaining organization, community, and environment.

Many drivers compel environmental resource management to take sustainability issues into account. Today's economic paradigms do not protect the natural environment, yet they deepen human dependency on biodiversity and ecosystem services. Ecologically, massive environmental degradation and climate change threaten the stability of ecological systems that humanity depends on. Socially, an increasing gap between rich and poor and the global North-South divide denies many access to basic human needs, rights, and education, leading to further environmental destruction. The planet's unstable condition is caused by many anthropogenic sources. As an exceptionally powerful contributing factor to social and environmental change, the modern organisation has the potential to apply environmental resource management with sustainability principals to achieve highly effective outcomes. To achieve sustainable development with environmental resource management an organisation should coincide with sustainability principles, such as: social and environmental accountability, long-term planning; a strong, shared vision; a holistic focus; devolved and consensus decision making; broad stakeholder engagement and justice; transparency measures; trust; and flexibility, to name a few.

Current Paradigm Shifts

To adjust to today's environment of quick social and ecological changes, some organizations have begun to experiment with various new tools and concepts. Those that are more traditional and stick to hierarchical decision making have difficulty dealing with the demand for lateral decision making that supports effective participation. Whether it be a matter of ethics or just strategic advantage organizations are internalizing sustainability principles. Examples of some of the world's largest and most profitable corporations who are shifting to sustainable environmental resource management are: Ford, Toyota, BMW, Honda, Shell, Du Pont, Swiss Re, Hewlett-Packard, and Unilever. An extensive study by the Boston Consulting Group reaching 1,560 business leaders from diverse regions, job positions, expertise in sustainability, industries, and sizes of organizations, revealed the many benefits of sustainable practice as well as its viability.

It is important to note that though sustainability of environmental resource management has improved, corporate sustainability, for one, has yet to reach the majority of global companies operating in the markets. The three major barriers to preventing organizations to shift towards sustainable practice with environmental resource management are: not understanding what sustainability is; having difficulty modeling an economically viable case for the switch; and having a flawed execution plan, or a lack thereof. Therefore, the most important part of shifting an organization to adopt sustainability in environmental resource management would be to create a shared vision and understanding of what sustainability is for that particular organization, and to clarify the business case.

Stakeholders

Public Sector

A conservation project in North Carolina involving the search for bog turtles was conducted by United States Fish and Wildlife Service and the North Carolina Wildlife Resources Commission and its volunteers

The public sector comprises the general government sector plus all public corporations including the central bank. In environmental resource management the public sector is responsible for administering natural resource management and implementing environmental protection legislation. The traditional role of the public sector in environmental resource management is to provide professional judgement through skilled technicians on behalf of the public. With the increase of intractable environmental problems, the public sector has been led to examine alternative paradigms for managing environmental resources. This has resulted in the public sector working collaboratively with other sectors (including other governments, private and civil) to encourage sustainable natural resource management behaviours.

Private Sector

The private sector comprises private corporations and non-profit institutions serving households. The private sector's traditional role in environmental resource management is that of the recovery of natural resources. Such private sector recovery groups include mining (minerals and petroleum), forestry and fishery organisations. Environmental resource management undertaken by the private sectors varies dependent upon the resource type, that being renewable or non-renewable and private and common resources (also see Tragedy of the Commons). Environmental managers from the private sector also need skills to manage collaboration within a dynamic social and political environment.

Civil Society

Civil society comprises associations in which societies voluntarily organise themselves into and which represent a wide range of interests and ties. These can include community-based organisations, indigenous peoples' organisations and non-government organisations (NGO). Functioning through strong public pressure, civil society can exercise their legal rights against the implementation of resource management plans, particularly land management plans. The aim of civil society in environmental resource management is to be included in the decision-making process by means of public participation. Public participation can be an effective strategy to invoke a sense of social responsibility of natural resources.

Tools

As with all management functions, effective management tools, standards and systems are required. An environmental management standard or system or protocol attempts to reduce environmental impact as measured by some objective criteria. The ISO 14001 standard is the most widely used standard for environmental risk management and is closely aligned to the European Eco-Management and Audit Scheme (EMAS). As a common auditing standard, the ISO 19011 standard explains how to combine this with quality management.

Other environmental management systems (EMS) tend to be based on the ISO 14001 standard and many extend it in various ways:

- The Green Dragon Environmental Management Standard is a five level EMS designed for smaller organisations for whom ISO 14001 may be too onerous and for larger organisations who wish to implement ISO 14001 in a more manageable step-by-step approach,

- BS 8555 is a phased standard that can help smaller companies move to ISO 14001 in six manageable steps,

- The Natural Step focuses on basic sustainability criteria and helps focus engineering on reducing use of materials or energy use that is unsustainable in the long term,

- Natural Capitalism advises using accounting reform and a general biomimicry and industrial ecology approach to do the same thing,

- US Environmental Protection Agency has many further terms and standards that it defines as appropriate to large-scale EMS,

- The UN and World Bank has encouraged adopting a "natural capital" measurement and management framework,

- The European Union Eco-Management and Audit Scheme (EMAS).

Other strategies exist that rely on making simple distinctions rather than building top-down management "systems" using performance audits and full cost accounting. For instance, Ecological Intelligent Design divides products into consumables, service products or durables and unsaleables — toxic products that no one should buy, or in many cases, do not realize they are buying. By eliminating the unsaleables from the comprehensive outcome of any purchase, better environmental resource management is achieved without *systems*.

Recent successful cases have put forward the notion of *integrated management*. It shares a wider approach and stresses out the importance of interdisciplinary assessment. It is an interesting notion that might not be adaptable to all cases.

Natural Resource Management

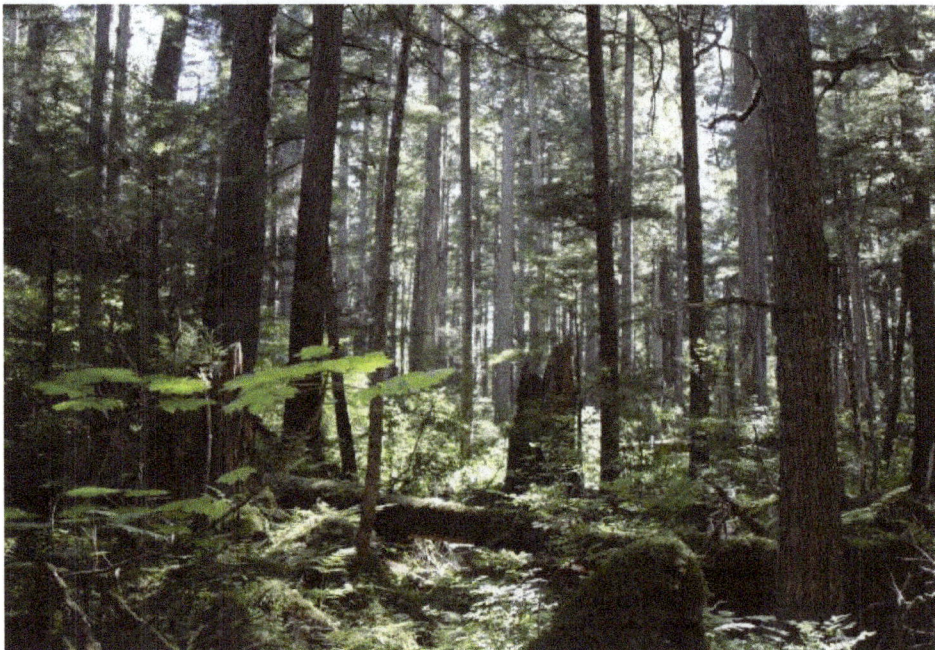

The Tongass National Forest in Alaska is managed by the United States Forest Service

Natural resource management refers to the management of natural resources such as land, water, soil, plants and animals, with a particular focus on how management affects the quality of life for both present and future generations (stewardship).

Natural resource management deals with managing the way in which people and natural land-scapes interact. It brings together land use planning, water management, biodiversity conservation, and the future sustainability of industries like agriculture, mining, tourism, fisheries and forestry. It recognises that people and their livelihoods rely on the health and productivity of our landscapes, and their actions as stewards of the land play a critical role in maintaining this health and productivity.

Natural resource management specifically focuses on a scientific and technical understanding of resources and ecology and the life-supporting capacity of those resources. Environmental management is also similar to natural resource management. In academic contexts, the sociology of natural resources is closely related to, but distinct from, natural resource management.

History

The Bureau of Land Management in the United States manages America's public lands, totaling approximately 264 million acres (1,070,000 km2) or one-eighth of the landmass of the country.

The emphasis on sustainability can be traced back to early attempts to understand the ecological nature of North American rangelands in the late 19th century, and the resource conservation movement of the same time. This type of analysis coalesced in the 20th century with recognition that preservationist conservation strategies had not been effective in halting the decline of natural resources. A more integrated approach was implemented recognising the intertwined social, cultural, economic and political aspects of resource management. A more holistic, national and even global form evolved, from the Brundtland Commission and the advocacy of sustainable development.

In 2005 the government of New South Wales, established a *Standard for Quality Natural Resource Management*, to improve the consistency of practice, based on an adaptive management approach.

In the United States, the most active areas of natural resource management are wildlife management often associated with ecotourism and rangeland management. In Australia, water sharing, such as the Murray Darling Basin Plan and catchment management are also significant.

Ownership Regimes

Natural resource management approaches can be categorised according to the kind and right of stakeholders, natural resources:

- State property
- Private property
- Common property
- Non-property (open access)
- Hybrid

State Property Regime

Ownership and control over the use of resources is in hands of the state. Individuals or groups may be able to make use of the resources, but only at the permission of the state. National forest, National parks and military reservations are some US examples.

Private Property Regime

Any property owned by a defined individual or corporate entity. Both the benefit and duties to the resources fall to the owner(s). Private land is the most common example.

Common Property Regimes

It is a private property of a group. The group may vary in size, nature and internal structure e.g. indigenous neighbours of village. Some examples of common property are community forests.

Non-Property Regimes (Open Access)

There is no definite owner of these properties. Each potential user has equal ability to use it as they wish. These areas are the most exploited. It is said that "Everybody's property is nobody's property". An example is a lake fishery. Common land may exist without ownership, in which case in the UK it is vested in a local authority.

Hybrid Regimes

Many ownership regimes governing natural resources will contain parts of more than one of the regimes described above, so natural resource managers need to consider the impact of hybrid regimes. An example of such a hybrid is native vegetation management in NSW, Australia, where legislation recognises a public interest in the preservation of native vegetation, but where most native vegetation exists on private land.

Stakeholder Analysis

Stakeholder analysis originated from business management practices and has been incorporated into natural resource management in ever growing popularity. Stakeholder analysis in the context of natural resource management identifies distinctive interest groups affected in the utilisation and conservation of natural resources.

There is no definitive definition of a stakeholder as illustrated in the table below. Especially in natural resource management as it is difficult to determine who has a stake and this will differ according to each potential stakeholder.

Different Approaches to Who is a Stakeholder:

Source	Who is a stakeholder	Kind of research
Freeman.	"can affect or is affected by the achievement of the organization's objectives"	Business Management
Bowie	"without whose support the organization would cease to exist"	Business Management
Clarkson	"...persons or groups that have, or claim, ownership, rights, or interests in a corporation and its activities, past, present, or future."	Business Management
Grimble and Wellard	"...any group of people, organized or unorganized, who share a common interest or stake in a particular issue or system..."	Natural resource management
Gass et al.	"...any individual, group and institution who would potentially be affected, whether positively or negatively, by a specified event, process or change."	Natural resource management
Buanes et al	"...any group or individual who may directly or indirectly affect—or be affected—...planning to be at least potential stakeholders."	Natural resource management
Brugha and Varvasovszky	"...actors who have an interest in the issue under consideration, who are affected by the issue, or who—because of their position—have or could have an active or passive influence on the decision making and implementation process."	Health policy
ODA	"... persons, groups or institutions with interests in a project or programme."	Development

Therefore, it is dependent upon the circumstances of the stakeholders involved with natural resource as to which definition and subsequent theory is utilised.

Billgrena and Holme identified the aims of stakeholder analysis in natural resource management:

- Identify and categorise the stakeholders that may have influence

- Develop an understanding of why changes occur

- Establish who can make changes happen

- How to best manage natural resources

This gives transparency and clarity to policy making allowing stakeholders to recognise conflicts of interest and facilitate resolutions. There are numerous stakeholder theories such as Mitchell et al. however Grimble created a framework of stages for a Stakeholder Analysis in natural resource management. Grimble designed this framework to ensure that the analysis is specific to the essential aspects of natural resource management.

Stages in Stakeholder Analysis:

- Clarify objectives of the analysis

- Place issues in a systems context

- Identify decision-makers and stakeholders

- Investigate stakeholder interests and agendas

- Investigate patterns of inter-action and dependence (e.g. conflicts and compatibilities, trade-offs and synergies)

Application:

Grimble and Wellard established that Stakeholder analysis in natural resource management is most relevant where issued can be characterised as;

- Cross-cutting systems and stakeholder interests

- Multiple uses and users of the resource.

- Market failure

- Subtractability and temporal trade-offs

- Unclear or open-access property rights

- Untraded products and services

- Poverty and under-representation

Case Studies:

In the case of the Bwindi Impenetrable National Park, a comprehensive stakeholder analysis would have been relevant and the Batwa people would have potentially been acknowledged as stakeholders preventing the loss of people's livelihoods and loss of life.

Nepal, Indonesia and Koreas' community forestry are successful examples of how stakeholder analysis can be incorporated into the management of natural resources. This allowed the stakeholders to identify their needs and level of involvement with the forests.

Criticisms:

- Natural resource management stakeholder analysis tends to include too many stakeholders which can create problems in of its self as suggested by Clarkson. "Stakeholder theory should not be used to weave a basket big enough to hold the world's misery."

- Starik proposed that nature needs to be represented as stakeholder. However this has been rejected by many scholars as it would be difficult to find appropriate representation and this representation could also be disputed by other stakeholders causing further issues.

- Stakeholder analysis can be used exploited and abused in order to marginalise other stakeholders.
- Identifying the relevant stakeholders for participatory processes is complex as certain stakeholder groups may have been excluded from previous decisions.
- On-going conflicts and lack of trust between stakeholders can prevent compromise and resolutions.

Alternatives/ Complementary Forms of Analysis:

- Social network analysis
- Common pool resource

Management Approaches

Natural resource management issues are inherently complex as they involve the ecological cycles, hydrological cycles, climate, animals, plants and geography etc. All these are dynamic and inter-related. A change in one of them may have far reaching and/or long term impacts which may even be irreversible. In addition to the natural systems, natural resource management also has to manage various stakeholders and their interests, policies, politics, geographical boundaries, economic implications and the list goes on. It is very difficult to satisfy all aspects at the same time. This results in conflicting situations.

After the United Nations Conference for the Environment and Development (UNCED) held in Rio de Janeiro in 1992, most nations subscribed to new principles for the integrated management of land, water, and forests. Although program names vary from nation to nation, all express similar aims.

The various approaches applied to natural resource management include:

- Top-down (command and control)
- Community-based natural resource management
- Adaptive management
- Precautionary approach
- Integrated natural resource management

Community-Based Natural Resource Management

The community-based natural resource management (CBNRM) approach combines conservation objectives with the generation of economic benefits for rural communities. The three key assumptions being that: locals are better placed to conserve natural resources, people will conserve a resource only if benefits exceed the costs of conservation, and people will conserve a resource that is linked directly to their quality of life. When a local people's quality of life is enhanced, their efforts and commitment to ensure the future well-being of the resource are also enhanced. Regional and community based natural resource management is also based on the principle of subsidiarity.

The United Nations advocates CBNRM in the Convention on Biodiversity and the Convention to Combat Desertification. Unless clearly defined, decentralised NRM can result an ambiguous socio-legal environment with local communities racing to exploit natural resources while they can e.g. forest communities in central Kalimantan (Indonesia).

A problem of CBNRM is the difficulty of reconciling and harmonising the objectives of socioeconomic development, biodiversity protection and sustainable resource utilisation. The concept and conflicting interests of CBNRM, show how the motives behind the participation are differentiated as either people-centred (active or participatory results that are truly empowering) or planner-centred (nominal and results in passive recipients). Understanding power relations is crucial to the success of community based NRM. Locals may be reluctant to challenge government recommendations for fear of losing promised benefits.

CBNRM is based particularly on advocacy by nongovernmental organizations working with local groups and communities, on the one hand, and national and transnational organizations, on the other, to build and extend new versions of environmental and social advocacy that link social justice and environmental management agendas with both direct and indirect benefits observed including a share of revenues, employment, diversification of livelihoods and increased pride and identity. CBNRM has raised new challenges, as concepts of community, territory, conservation, and indigenous are worked into politically varied plans and programs in disparate sites. Warner and Jones address strategies for effectively managing conflict in CBNRM.

The capacity of indigenous communities to conserve natural resources has been acknowledged by the Australian Government with the Caring for Country Program. Caring for our Country is an Australian Government initiative jointly administered by the Australian Government Department of Agriculture, Fisheries and Forestry and the Department of the Environment, Water, Heritage and the Arts. These Departments share responsibility for delivery of the Australian Government's environment and sustainable agriculture programs, which have traditionally been broadly referred to under the banner of 'natural resource management'.

These programs have been delivered regionally, through 56 State government bodies, successfully allowing regional communities to decide the natural resource priorities for their regions.

Governance is seen as a key consideration for delivering community-based or regional natural resource management. In the State of NSW, the 13 catchment management authorities (CMAs) are overseen by the Natural Resources Commission (NRC), responsible for undertaking audits of the effectiveness of regional natural resource management programs.

Adaptive Management

The primary methodological approach adopted by catchment management authorities (CMAs) for regional natural resource management in Australia is adaptive management.

This approach includes recognition that adaption occurs through a process of 'plan-do-review-act'. It also recognises seven key components that should be considered for quality natural resource management practice:

- Determination of scale

- Collection and use of knowledge

- Information management

- Monitoring and evaluation

- Risk management

- Community engagement

- Opportunities for collaboration.

Integrated Natural Resource Management

Integrated natural resource management (INRM) is a process of managing natural resources in a systematic way, which includes multiple aspects of natural resource use (biophysical, socio-political, and economic) meet production goals of producers and other direct users (e.g., food security, profitability, risk aversion) as well as goals of the wider community (e.g., poverty alleviation, welfare of future generations, environmental conservation). It focuses on sustainability and at the same time tries to incorporate all possible stakeholders from the planning level itself, reducing possible future conflicts. The conceptual basis of INRM has evolved in recent years through the convergence of research in diverse areas such as sustainable land use, participatory planning, integrated watershed management, and adaptive management. INRM is being used extensively and been successful in regional and community based natural management.

Frameworks and Modelling

There are various frameworks and computer models developed to assist natural resource management.

Geographic Information Systems (GIS)

GIS is a powerful analytical tool as it is capable of overlaying datasets to identify links. A bush regeneration scheme can be informed by the overlay of rainfall, cleared land and erosion. In Australia, Metadata Directories such as NDAR provide data on Australian natural resources such as vegetation, fisheries, soils and water. These are limited by the potential for subjective input and data manipulation.

Natural Resources Management Audit Frameworks

The NSW Government in Australia has published an audit framework for natural resource management, to assist the establishment of a performance audit role in the governance of regional natural resource management. This audit framework builds from other established audit methodologies, including performance audit, environmental audit and internal audit. Audits undertaken using this framework have provided confidence to stakeholders, identified areas for improvement and described policy expectations for the general public.

The Australian Government has established a framework for auditing greenhouse emissions and energy reporting, which closely follows Australian Standards for Assurance Engagements.

The Australian Government is also currently preparing an audit framework for auditing water management, focussing on the implementation of the Murray Darling Basin Plan.

Other Elements

Biodiversity Conservation

The issue of biodiversity conservation is regarded as an important element in natural resource management. What is biodiversity? Biodiversity is a comprehensive concept, which is a description of the extent of natural diversity. Gaston and Spicer (p. 3) point out that biodiversity is "the variety of life" and relate to different kinds of "biodiversity organization". According to Gray (p. 154), the first widespread use of the definition of biodiversity, was put forward by the United Nations in 1992, involving different aspects of biological diversity.

Precautionary Biodiversity Management

The "threats" wreaking havoc on biodiversity include; habitat fragmentation, putting a strain on the already stretched biological resources; forest deterioration and deforestation; the invasion of "alien species" and "climate change"(p. 2). Since these threats have received increasing attention from environmentalists and the public, the precautionary management of biodiversity becomes an important part of natural resources management. According to Cooney, there are material measures to carry out precautionary management of biodiversity in natural resource management.

Concrete "Policy Tools"

Cooney claims that the policy making is dependent on "evidences", relating to "high standard of proof", the forbidding of special "activities" and "information and monitoring requirements". Before making the policy of precaution, categorical evidence is needed. When the potential menace of "activities" is regarded as a critical and "irreversible" endangerment, these "activities" should be forbidden. For example, since explosives and toxicants will have serious consequences to endanger human and natural environment, the South Africa Marine Living Resources Act promulgated a series of policies on completely forbidding to "catch fish" by using explosives and toxicants.

Administration and Guidelines

According to Cooney, there are 4 methods to manage the precaution of biodiversity in natural resources management;

1. "Ecosystem based Management" including "more risk-averse and precautionary management", where "given prevailing uncertainty regarding ecosystem structure, function, and inter-specific interactions, precaution demands an ecosystem rather than single-species approach to management".

2. "Adaptive management" is "a management approach that expressly tackles the uncertainty and dynamism of complex systems".

3. "Environmental impact assessment" and exposure ratings decrease the "uncertainties" of precaution, even though it has deficiencies, and

4. "Protectionist approaches", which "most frequently links to" biodiversity conservation in natural resources management.

Land Management

In order to have a sustainable environment, understanding and using appropriate management strategies is important. In terms of understanding, Young emphasises some important points of land management:

- Comprehending the processes of nature including ecosystem, water, soils

- Using appropriate and adapting management systems in local situations

- Cooperation between scientists who have knowledge and resources and local people who have knowledge and skills

Dale et al. (2000) study has shown that there are five fundamental and helpful ecological principles for the land manager and people who need them. The ecological principles relate to time, place, species, disturbance and the landscape and they interact in many ways.It is suggested that land managers could follow these guidelines:

- Examine impacts of local decisions in a regional context, and the effects on natural resources.

- Plan for long-term change and unexpected events.

- Preserve rare landscape elements and associated species.

- Avoid land uses that deplete natural resources.

- Retain large contiguous or connected areas that contain critical habitats.

- Minimize the introduction and spread of non-native species.

- Avoid or compensate for the effects of development on ecological processes.

- Implement land-use and land-management practices that are compatible with the natural potential of the area.

Fisheries Management

Fisheries management draws on fisheries science in order to find ways to protect fishery resources so sustainable exploitation is possible. Modern fisheries management is often referred to as a governmental system of appropriate management rules based on defined objectives and a mix of management means to implement the rules, which are put in place by a system of monitoring control and surveillance. According to the Food and Agriculture Organization of the United Nations (FAO), there are "no clear and generally accepted definitions of fisheries management". However, the working definition used by the FAO and much cited elsewhere is:

The integrated process of information gathering, analysis, planning, consultation, decision-making, allocation of resources and formulation and implementation, with enforcement as necessary, of regulations or rules which govern fisheries activities in order to ensure the continued productivity of the resources and the accomplishment of other fisheries objectives.

History

Fisheries have been explicitly managed in some places for hundreds of years. More than 80 percent of the worlds commercial exploitation of fish and shellfish are harvest from natural occurring populations in the oceans and freshwater areas. For example, the Māori people, New Zealand residents for about 700 years, had prohibitions against taking more than what could be eaten and about giving back the first fish caught as an offering to sea god Tangaroa. Starting in the 18th century attempts were made to regulate fishing in the North Norwegian fishery. This resulted in the enactment of a law in 1816 on the Lofoten fishery, which established in some measure what has come to be known as territorial use rights.

"The fishing banks were divided into areas belonging to the nearest fishing base on land and further subdivided into fields where the boats were allowed to fish. The allocation of the fishing fields was in the hands of local governing committees, usually headed by the owner of the onshore facilities which the fishermen had to rent for accommodation and for drying the fish."

Governmental resource protection-based fisheries management is a relatively new idea, first developed for North European fisheries after the first Overfishing Conference held in London in 1936. In 1957 British fisheries researchers Ray Beverton and Sidney Holt published a seminal work on North Sea commercial fisheries dynamics. In the 1960s the work became the theoretical platform for North European management schemes.

After some years away from the field of fisheries management, Beverton criticized his earlier work in a paper given at the first World Fisheries Congress in Athens in 1992. "The Dynamics of Exploited Fish Populations" expressed his concerns, including the way his and Sidney Holt's work had been misinterpreted and misused by fishery biologists and managers during the previous 30 years. Nevertheless, the institutional foundation for modern fishery management had been laid.

In 1996, the Marine Stewardship Council was founded to set standards for sustainable fishing. In 2010, the Aquaculture Stewardship Council was created to do the same for aquaculture.

A report by Prince Charles' International Sustainability Unit, the New York-based Environmental Defense Fund and 50in10 published in July 2014 estimated global fisheries were adding $270 billion a year to global GDP, but by full implementation of sustainable fishing, that figure could rise by an extra amount of as much as $50 billion.

Political Objectives

According to the FAO, fisheries management should be based explicitly on political objectives, ideally with transparent priorities. Typical political objectives when exploiting a fish resource are to:

- maximize sustainable biomass yield
- maximize sustainable economic yield
- secure and increase employment
- secure protein production and food supplies
- increase export income

Such political goals can also be a weak part of fisheries management, since the objectives can conflict with each other.

International Objectives

Fisheries objectives need to be expressed in concrete management rules. In most countries fisheries management rules should be based on the internationally agreed, though non-binding, Code of Conduct for Responsible Fisheries, agreed at a meeting of the U.N.'s Food and Agriculture Organization FAO session in 1995. The precautionary approach it prescribes is typically implemented in concrete management rules as minimum spawning biomass, maximum fishing mortality rates, etc. In 2005 the UBC Fisheries Centre at the University of British Columbia comprehensively reviewed the performance of the world's major fishing nations against the Code.

International agreements are required in order to regulate fisheries in international waters. The desire for agreement on this and other maritime issues led to three conferences on the Law of the Sea, and ultimately to the treaty known as the United Nations Convention on the Law of the Sea (UNCLOS). Concepts such as exclusive economic zones (EEZ, extending 200 nautical miles (370 km) from a nation's coasts) allocate certain sovereign rights and responsibilities for resource management to individual countries.

Other situations need additional intergovernmental coordination. For example, in the Mediterranean Sea and other relatively narrow bodies of water, EEZ of 200 nautical miles (370 km) are irrelevant. International waters beyond 12-nautical-mile (22 km) from shore require explicit agreements.

Straddling fish stocks, which migrate through more than one EEZ also present challenges. Here sovereign responsibility must be agreed with neighbouring coastal states and fishing entities. Usually this is done through the medium of a regional organisation set up for the purpose of coordinating the management of that stock.

UNCLOS does not prescribe precisely how fisheries confined only to international waters should be managed. Several new fisheries (such as high seas bottom trawling fisheries) are not (yet) subject to international agreement across their entire range. In November 2004 the UN General Assembly issued a resolution on Fisheries that prepared for further development of international fisheries management law.

Management Mechanisms

Many countries have set up Ministries/Government Departments, named "Ministry of Fisheries" or similar, controlling aspects of fisheries within their exclusive economic zones. Four categories of management means have been devised, regulating either input/investment, or output, and operating either directly or indirectly:

	Inputs	Outputs
Indirect	Vessel licensing	Catching techniques
Direct	Limited entry	Catch quota and technical regulation

Technical means may include:

- prohibiting devices such as bows and arrows, and spears, or firearms

- prohibiting nets

- setting minimum mesh sizes

- limiting the average potential catch of a vessel in the fleet (vessel and crew size, gear, electronic gear and other physical "inputs".

- prohibiting bait

- snagging

- limits on fish traps

- limiting the number of poles or lines per fisherman

- restricting the number of simultaneous fishing vessels

- limiting a vessel's average operational intensity per unit time at sea

- limiting average time at sea

Catch Quotas

Systems that use *individual transferable quotas* (ITQ), also called individual fishing quota limit the total catch and allocate shares of that quota among the fishers who work that fishery. Fishers can buy/sell/trade shares as they choose.

A large scale study in 2008 provided strong evidence that ITQ's can help to prevent fishery collapse and even restore fisheries that appear to be in decline. Other studies have shown negative socio-economic consequences of ITQs, especially on small-sclale fisheries. These consequences include concentration of quota in that hands of few fishers; increased number of inactive fishers leasing their quotas to others (a phenomenon known as armchair fishermen); and detrimental effects on coastal communities.

Precautionary Principle

The *Fishery Manager's Guidebook* issued in 2009 by the FAO of the United Nations, advises that the precautionary approach or principle should be applied when "ecosystem resilience and human impact (including reversibility) are difficult to forecast and hard to distinguish from natural changes." The precautionary principle suggests that when an action risks harm, it should not be proceeded with until it can be scientifically proven to be safe. Historically fishery managers have applied this principle the other way round; fishing activities have not been curtailed until it has been proven that they have already damaged existing ecosystems. In a paper published in 2007, Shertzer and Prager suggested that there can be significant benefits to stock biomass and fishery yield if management is stricter and more prompt.

Fisheries Law

Fisheries law is an emerging and specialized area of law which includes the study and analysis of different fisheries management approaches, including seafood safety regulations and aquaculture regulations. Despite its importance, this area is rarely taught at law schools around the world, which leaves a vacuum of advocacy and research.

Climate Change

In the past, changing climate has affected inland and offshore fisheries and such changes are likely to continue. From a fisheries perspective, the specific driving factors of climate change include rising water temperature, alterations in the hydrologic cycle, changes in nutrient fluxes, and relocation of spawning and nursery habitat. Further, changes in such factors would affect resources at all levels of biological organization, including the genetic, organism, population, and ecosystem levels.

Population Dynamics

Population dynamics describes the growth and decline of a given fishery stock over time, as controlled by birth, death and migration. It is the basis for understanding changing fishery patterns and issues such as habitat destruction, predation and optimal harvesting rates. The population dynamics of fisheries has been traditionally used by fisheries scientists to determine sustainable yields.

The basic accounting relation for population dynamics is the BIDE model:

$$N_1 = N_0 + B - D + I - E$$

where N_1 is the number of individuals at time 1, N_0 is the number of individuals at time 0, B is the number of individuals born, D the number that died, I the number that immigrated, and E the number that emigrated between time 0 and time 1. While immigration and emigration can be present in wild fisheries, they are usually not measured.

Care is needed when applying population dynamics to real world fisheries. In the past, over-simplistic modelling, such as ignoring the size, age and reproductive status of the fish, focusing solely on a single species, ignoring bycatch and physical damage to the ecosystem, has accelerated the collapse of key stocks.

Ecosystem Based Fisheries

> 66
>
> *We propose that rebuilding ecosystems, and not sustainability per se, should be the goal of fishery management. Sustainability is a deceptive goal because human harvesting of fish leads to a progressive simplification of ecosystems in favour of smaller, high turnover, lower trophic level fish species that are adapted to withstand disturbance and habitat degradation.*
>
> 99
>
> — *Tony Pitcher and Daniel Pauly,*

According to marine ecologist Chris Frid, the fishing industry points to pollution and global warming as the causes of unprecedentedly low fish stocks in recent years, writing, "Everybody would like to see the rebuilding of fish stocks and this can only be achieved if we understand all of the influences, human and natural, on fish dynamics." Overfishing has also had an effect. Frid adds, "Fish communities can be altered in a number of ways, for example they can decrease if particular sized individuals of a species are targeted, as this affects predator and prey dynamics. Fishing, however, is not the sole perpetrator of changes to marine life - pollution is another example No one factor operates in isolation and components of the ecosystem respond differently to each individual factor."

In contrast to the traditional approach of focusing on a single species, the ecosystem-based approach is organized in terms of ecosystem services. Ecosystem-based fishery concepts have been implemented in some regions. In 2007 a group of scientists offered the following *ten commandments*

> - Keep a perspective that is holistic, risk-adverse and adaptive.
>
> - Maintain an "old growth" structure in fish populations, since big, old and fat female fish have been shown to be the best spawners, but are also susceptible to overfishing.
>
> - Characterize and maintain the natural spatial structure of fish stocks, so that management boundaries match natural boundaries in the sea.
>
> - Monitor and maintain seafloor habitats to make sure fish have food and shelter.
>
> - Maintain resilient ecosystems that are able to withstand occasional shocks.
>
> - Identify and maintain critical food-web connections, including predators and forage species.
>
> - Adapt to ecosystem changes through time, both short-term and on longer cycles of decades or centuries, including global climate change.
>
> - Account for evolutionary changes caused by fishing, which tends to remove large, older fish.
>
> - Include the actions of humans and their social and economic systems in all ecological equations.

Report to Congress (2009): The State of Science to Support an Ecosystem Approach to Regional Fishery Management National Marine Fisheries Service, NOAA Technical Memorandum NMFS-F/SPO-96.

Elderly Maternal Fish

Traditional management practices aim to reduce the number of old, slow-growing fish, leaving more room and resources for younger, faster-growing fish. Most marine fish produce huge numbers of eggs. The assumption was that younger spawners would produce plenty of viable larvae.

Old fat female rockfish are the best producers

However, 2005 research on rockfish shows that large, elderly females are far more important than younger fish in maintaining productive fisheries. The larvae produced by these older maternal fish grow faster, survive starvation better, and are much more likely to survive than the offspring of younger fish. Failure to account for the role of older fish may help explain recent collapses of some major US West Coast fisheries. Recovery of some stocks is expected to take decades. One way to prevent such collapses is to establish marine reserves, where fishing is not allowed and fish populations age naturally.

Data Quality

According to fisheries scientist Milo Adkison, the primary limitation in fisheries management decisions is the absence of quality data. Fisheries management decisions are often based on population models, but the models need quality data to be effective. He asserts that scientists and fishery managers would be better served with simpler models and improved data.

The most reliable source for summary statistics is the FAO Fisheries Department.

Ecopath

Ecopath, with Ecosim (EwE), is an ecosystem modelling software suite. It was initially a NOAA initiative led by Jeffrey Polovina, later primarily developed at the UBC Fisheries Centre of the University of British Columbia. In 2007, it was named as one of the ten biggest scientific breakthroughs in NOAA's 200-year history. The citation states that Ecopath "revolutionized scientists' ability worldwide to understand complex marine ecosystems". Behind this lies two decades of development work by Villy Christensen, Carl Walters, Daniel Pauly, and other fisheries scientists. As of 2010 there are 6000 registered users in 155 countries. Ecopath is widely used in fisheries management as a tool for modelling and visualising the complex relationships that exist in real world marine ecosystems.

Human Factors

Managing fisheries is about managing people and businesses, and not about managing fish. Fish populations are managed by regulating the actions of people. If fisheries management is to be successful, then associated human factors, such as the reactions of fishermen, are of key importance, and need to be understood.

Management regulations must also consider the implications for stakeholders. Commercial fishermen rely on catches to provide for their families just as farmers rely on crops. Commercial fishing can be a traditional trade passed down from generation to generation. Most commercial fishing is based in towns built around the fishing industry; regulation changes can impact an entire town's economy. Cuts in harvest quotas can have adverse effects on the ability of fishermen to compete with the tourism industry.

Performance

The biomass of global fish stocks has been allowed to run down. This biomass is now diminished to the point where it is no longer possible to sustainably catch the amount of fish that could be caught. According to a 2008 UN report, titled *The Sunken Billions: The Economic Justification for Fisheries Reform*, the world's fishing fleets incur a "$US 50 billion annual economic loss" through depleted stocks and poor fisheries management. The report, produced jointly by the World Bank and the UN Food and Agriculture Organization (FAO), asserts that half the world's fishing fleet could be scrapped with no change in catch.

"By improving governance of marine fisheries, society could capture a substantial part of this $50 billion annual economic loss. Through comprehensive reform, the fisheries sector could become a basis for economic growth and the creation of alternative livelihoods in many countries. At the same time, a nation's natural capital in the form of fish stocks could be greatly increased and the negative impacts of the fisheries on the marine environment reduced."

The most prominent failure of fisheries management in recent times has perhaps been the events that lead to the collapse of the northern cod fisheries. More recently, the International Consortium of Investigative Journalists produced a series of journalistic investigations called *Looting the seas*. These detail investigations into the black market for bluefin tuna, the subsidies propping up the Spanish fishing industry, and the overfishing of the Chilean jack mackerel.

Reference

- Berman, Morris (1981). The Reenchantment of the World. Cornell University Press. ISBN 978-0-8014-9225-9. Retrieved 31 October 2013.

- Pepper, David; Perkins, John W.; Youngs, Martyn J. (1984). The Roots of Modern Environmentalism. Croom Helm. ISBN 978-0-7099-2064-9.

- Thampapillai, Dodo J. (2002). Environmental economics: concepts, methods, and policies. Oxford University Press. ISBN 978-0-19-553577-8.

- Daly, Herman E.; Cobb, John B. Jr (1994). For The Common Good: Redirecting the Economy toward Community, the Environment, and a Sustainable Future. Beacon Press. ISBN 978-0-8070-4705-7.

- Walters, Carl J. (1986). Adaptive management of renewable resources. Macmillan. ISBN 978-0-02-947970-4.

- United Nations Environment Programme (1978). Holling, C.S., ed. Adaptive environmental assessment and management. International Institute for Applied Systems Analysis. ISBN 978-0-471-99632-3.

- Gunderson, Lance H.; Holling, C.S., eds. (2002). Panarchy Synopsis: Understanding Transformations in Human and Natural Systems. Island Press. ISBN 978-1-55963-330-7.

- Armitage, D.R.; Berkes, F.; Doubleday, N. (2007). Adaptive Co-Management: Collaboration, Learning, and Multi-Level Governance. Vancouver: UBC Press. ISBN 978-0-7748-1383-9

- Avery, Gayle C.; Bergsteiner, Harald (2010). Honeybees and Locusts: The Business Case for Sustainable Leadership. Allen & Unwin. ISBN 978-1-74237-393-5.

- Costanza, Robert; Norton, Bryan G.; Haskell, Benjamin D. (1992). Ecosystem Health: New Goals for Environmental Management. Island Press. ISBN 978-1-55963-140-2.

- *Arnason, R; Kelleher, K; Willmann, R (2008). The Sunken Billions: The Economic Justification for Fisheries Reform. World Bank and FAO. ISBN 978-0-8213-7790-1.*

- *Beverton, R. J. H.; Holt, S. J. (1957). On the Dynamics of Exploited Fish Populations. Fishery Investigations Series II Volume XIX. Chapman and Hall (Blackburn Press, 2004). ISBN 978-1-930665-94-1.*

- Caddy JF and Mahon R (1995) "Reference points for fisheries management" FAO Fisheries technical paper 347, Rome. ISBN 92-5-103733-7

- McGoodwin JR (2001) *Understanding the cultures of fishing communities. A key to fisheries management and food security* FAO Fisheries, Technical Paper 401. ISBN 978-92-5-104606-7.

- Morgan, Gary; Staples, Derek and Funge-Smith, Simon (2007) *Fishing capacity management and illegal, unreported and unregulated fishing in Asia* FAO RAP Publication. 2007/17. ISBN 978-92-5-005669-2

- Townsend, R; Shotton, Ross and Uchida, H (2008) *Case studies in fisheries self-governance* FAO Fisheries Technical Paper. No 504. ISBN 978-92-5-105897-8

- Walters, Carl J. and Steven J. D. Martell (2004) *Fisheries ecology and management* Princeton University Press. ISBN 978-0-691-11545-0.

- *Grimble, R (1998). Stakeholder methodologies in natural resource management, Socioeconomic Methodologies (PDF). Chatham: Natural Resources Institute. pp. 1–12. Retrieved 27 October 2014.*

- *Kellert, S; Mehta, J; Ebbin, S; Litchtenfeld, L. (2000). Community natural resource management: promise, rhetoric, and reality (PDF). Society and Natural Resources, 13:705-715. Retrieved 27 October 2014.*

- *Brosius, J.; Peter Tsing; Anna Lowenhaupt; Zerner, Charles (1998). Representing communities: Histories and politics of community-based natural resource management. 11. Society & Natural Resources: An International Journal. pp. 157–168. Retrieved 27 October 2014.*

- *Lovell, C.; Mandondo A.; Moriarty P. (2002). The question of scale in integrated natural resource management. Conservation Ecology 5(2): 25. Retrieved 27 October 2014.*

- *Cooney, R (2004). The Precautionary Principle in Biodiversity Conservation and Natural Resource Management (PDF). IUCN Policy and Global Change Series. Retrieved 27 October 2014.*

Modern Concepts of Environmental Protection and Management

This chapter will discuss the modern concepts used in protecting and conserving the environment. The main topics discussed in this chapter are tragedy of commons, ecosystem services, habitat conservation, environmental management system and life-cycle assessment. This chapter is apt for learning the present concepts of this area.

Tragedy of the Commons

Cows on Selsley Common. The "tragedy of the commons" is one way of accounting for overexploitation.

The tragedy of the commons is an economic theory of a situation within a shared-resource system where individual users acting independently according to their own self-interest behave contrary to the common good of all users by depleting that resource through their collective action.

The concept and name originate in an essay written in 1833 by the Victorian economist William Forster Lloyd, who used a hypothetical example of the effects of unregulated grazing on common land (then colloquially called "the commons") in the British Isles. The concept became widely known over a century later due to an article written by the ecologist Garrett Hardin in 1968. In this context, commons is taken to mean any shared and unregulated resource such as atmosphere, oceans, rivers, fish stocks, or even an office refrigerator.

The tragedy of the commons is often cited in connection with sustainable development, meshing economic growth and environmental protection, as well as in the debate over global warming. It has also been used in analyzing behavior in the fields of economics, evolutionary psychology, anthropology, game theory, politics, taxation and sociology.

Although commons have been known to collapse due to overuse (such as in over-fishing), many examples exist where communities use common resources prudently without collapse. According to the political economist Elinor Ostrom, although it is often claimed that only private ownership or government regulation can prevent the "tragedy of the commons", it is in the interests of the users of a commons to manage it prudently, and complex social schemes are often devised by them for maintaining common resources efficiently.

Expositions

Lloyd's Pamphlet

In 1833 the English economist William Forster Lloyd published a pamphlet which included a hypothetical example of over-use of a common resource. This was the situation of cattle herders sharing a common parcel of land on which they are each entitled to let their cows graze, as was the custom in English villages. He postulated that if a herder put more than his allotted number of cattle on the common, overgrazing could result. For each additional animal, a herder could receive additional benefits, but the whole group shared damage to the commons. If all herders made this individually rational economic decision, the common could be depleted or even destroyed, to the detriment of all.

Garrett Hardin's Article

In 1968, ecologist Garrett Hardin explored this social dilemma in his article "The Tragedy of the Commons", published in the journal *Science*. The essay derived its title from the pamphlet by Lloyd, which he cites, on the over-grazing of common land.

Hardin discussed problems that cannot be solved by technical means, as distinct from those with solutions that require "a change only in the techniques of the natural sciences, demanding little or nothing in the way of change in human values or ideas of morality". Hardin focused on human population growth, the use of the Earth's natural resources, and the welfare state. Hardin argued that if individuals relied on themselves alone, and not on the relationship of society and man, then the number of children had by each family would not be of public concern. Parents breeding excessively would leave fewer descendants because they would be unable to provide for each child adequately. Such negative feedback is found in the animal kingdom. Hardin said that if the children of improvident parents starved to death, if overbreeding was its own punishment, then there would be no public interest in controlling the breeding of families. Hardin blamed the welfare state for allowing the tragedy of the commons; where the state provides for children and supports overbreeding as a fundamental human right, Malthusian catastrophe is inevitable. Consequently, in his article, Hardin lamented the following proposal from the United Nations:

The Universal Declaration of Human Rights describes the family as the natural and fundamental unit of society. [Article 16] It follows that any choice and decision with regard to the size of the family must irrevocably rest with the family itself, and cannot be made by anyone else.

— U Thant, Statement on Population by UN Secretary-General

In addition, Hardin also pointed out the problem of individuals acting in rational self-interest by claiming that if all members in a group used common resources for their own gain and with no regard for others, all resources would still eventually be depleted. Overall, Hardin argued against relying on conscience as a means of policing commons, suggesting that this favors selfish individuals – often known as free riders – over those who are more altruistic.

In the context of avoiding over-exploitation of common resources, Hardin concluded by restating Hegel's maxim (which was quoted by Engels), "freedom is the recognition of necessity". He suggested that "freedom" completes the tragedy of the commons. By recognizing resources as commons in the first place, and by recognizing that, as such, they require management, Hardin believed that humans "can preserve and nurture other and more precious freedoms".

The "Commons" as a Modern Resource Concept

Hardin's article was the start of the modern use of "Commons" as a shared resource term. As Frank van Laerhoven & Elinor Ostrom have stated: "Prior to the publication of Hardin's article on the tragedy of the commons (1968), titles containing the words 'the commons', 'common pool resources,' or 'common property' were very rare in the academic literature." They go on to say; "In 2002, Barrett and Mabry conducted a major survey of biologists to determine which publications in the twentieth century had become classic books or benchmark publications in biology. They report that Hardin's 1968 article was the one having the greatest career impact on biologists and the most frequently cited".

Application

Metaphoric Meaning

Like Lloyd and Thomas Malthus before him, Hardin was primarily interested in the problem of human population growth. But in his essay, he also focused on the use of larger (though finite) resources such as the Earth's atmosphere and oceans, as well as pointing out the "negative commons" of pollution (i.e., instead of dealing with the deliberate privatization of a positive resource, a "negative commons" deals with the deliberate commonization of a negative cost, pollution).

As a metaphor, the tragedy of the commons should not be taken too literally. The "tragedy" is not in the word's conventional or theatric sense, nor a condemnation of the processes that lead to it. Similarly, Hardin's use of "commons" has frequently been misunderstood, leading him to later remark that he should have titled his work "The Tragedy of the Unregulated Commons".

The metaphor illustrates the argument that free access and unrestricted demand for a finite resource ultimately reduces the resource through over-exploitation, temporarily or permanently. This occurs because the benefits of exploitation accrue to individuals or groups, each of whom is motivated to maximize use of the resource to the point in which they become reliant on it, while the costs of the exploitation are borne by all those to whom the resource is available (which may be a wider class of individuals than those who are exploiting it). This, in turn, causes demand for the resource to increase, which causes the problem to snowball until the resource collapses (even if it retains a capacity to recover). The rate at which depletion of the resource is realized depends primarily on three factors: the number of users wanting to consume the common in question, the consumptiveness of their uses, and the relative robustness of the common.

The same concept is sometimes called the "tragedy of the fishers", because fishing too many fish before or during breeding could cause stocks to plummet.

Modern Commons

The *tragedy of the commons* can be considered in relation to environmental issues such as sustainability. The commons dilemma stands as a model for a great variety of resource problems in society today, such as water, forests, fish, and non-renewable energy sources such as oil and coal.

Situations exemplifying the "tragedy of the commons" include the overfishing and destruction of the Grand Banks, the destruction of salmon runs on rivers that have been dammed – most prominently in modern times on the Columbia River in the Northwest United States, and historically in North Atlantic rivers – the devastation of the sturgeon fishery – in modern Russia, but historically in the United States as well – and, in terms of water supply, the limited water available in arid regions (e.g., the area of the Aral Sea) and the Los Angeles water system supply, especially at Mono Lake and Owens Lake.

In economics, an externality is a cost or benefit that affects a party who did not choose to incur that cost or benefit. Negative externalities are a well-known feature of the "tragedy of the commons". For example, driving cars has many negative externalities; these include pollution, carbon emissions, and traffic accidents. Every time 'Person A' gets in a car, it becomes more likely that 'Person Z' – and millions of others – will suffer in each of those areas. Economists often urge the government to adopt policies that "internalize" an externality.

Examples

More general examples (some alluded to by Hardin) of potential and actual tragedies include:

Clearing rainforest for agriculture in southern Mexico.

- Planet Earth ecology
  - Uncontrolled human population growth leading to overpopulation.
  - A preference for sons made people abort baby girls. This results in an imbalanced gender ratio.
  - Air, whether ambient air polluted by industrial emissions and cars among other sources of air pollution, or indoor air
  - Water – Water pollution, water crisis of over-extraction of groundwater and wasting water due to overirrigation
  - Forests – Frontier logging of old growth forest and slash and burn
  - Energy resources and climate – Environmental residue of mining and drilling, Burning of fossil fuels and consequential global warming
  - Animals – Habitat destruction and poaching leading to the Holocene mass extinction
  - Oceans – Overfishing
  - Antibiotics – Antibiotic Resistance Mis-use of antibiotics anywhere in the world will eventually result in antibiotic resistance developing at an accelerated rate. The resulting antibiotic resistance has spread (and will likely continue to do so in the future) to other bacteria and other regions, hurting or destroying the Antibiotic Commons that is shared on a world-wide basis

- Publicly shared resources
  - Spam email degrades the usefulness of the email system and increases the cost for all users of the Internet while providing a benefit to only a tiny number of individuals.
  - Vandalism and littering in public spaces such as parks, recreation areas, and public restrooms.
  - Knowledge commons encompass immaterial and collectively owned goods in the information age.
    - Including, for example, source code and software documentation in software projects that can get "polluted" with messy code or inaccurate information.

Application to Evolutionary Biology

A parallel was drawn recently between the tragedy of the commons and the competing behaviour of parasites that through acting selfishly eventually diminish or destroy their common host.

The idea has also been applied to areas such as the evolution of virulence or sexual conflict, where males may fatally harm females when competing for matings. It is also raised as a question in studies of social insects, where scientists wish to understand why insect workers do not undermine the "common good" by laying eggs of their own and causing a breakdown of the society.

The idea of evolutionary suicide, where adaptation at the level of the individual causes the whole species or population to be driven extinct, can be seen as an extreme form of an evolutionary tragedy of the commons.

From an evolutionary point of view, the creation of the tragedy of the commons in pathogenic microbes may provide us with advanced therapeutic ways.

Commons Dilemma

The *commons dilemma* is a specific class of social dilemma in which people's short-term selfish interests are at odds with long-term group interests and the common good. In academia, a range of related terminology has also been used as shorthand for the theory or aspects of it, including *resource dilemma*, *take-some dilemma*, and *common pool resource*.

Commons dilemma researchers have studied conditions under which groups and communities are likely to under- or over-harvest common resources in both the laboratory and field. Research programs have concentrated on a number of motivational, strategic, and structural factors that might be conducive to management of commons.

In game theory, which constructs mathematical models for individuals' behavior in strategic situations, the corresponding "game", developed by Hardin, is known as the Commonize Costs – Privatize Profits Game (CC–PP game).

Psychological Factors

Kopelman, Weber, & Messick (2002), in a review of the experimental research on cooperation in commons dilemmas, identify nine classes of independent variables that influence cooperation in commons dilemmas: social motives, gender, payoff structure, uncertainty, power and status, group size, communication, causes, and frames. They organize these classes and distinguish between psychological individual differences (stable personality traits) and situational factors (the environment). Situational factors include both the task (social and decision structure) and the perception of the task.

Empirical findings support the theoretical argument that the cultural group is a critical factor that needs to be studied in the context of situational variables. Rather than behaving in line with economic incentives, people are likely to approach the decision to cooperate with an appropriateness framework. An expanded, four factor model of the Logic of Appropriateness, suggests that the cooperation is better explained by the question: "What does a person like me (identity) do (rules) in a situation like this (recognition) given this culture (group)?"

Strategic Factors

Strategic factors also matter in commons dilemmas. One often-studied strategic factor is the order in which people take harvests from the resource. In simultaneous play, all people harvest at the same time, whereas in sequential play people harvest from the pool according to a predetermined sequence – first, second, third, etc. There is a clear order effect in the latter games: the harvests of those who come first – the leaders – are higher than the harvest of those coming later – the followers. The interpretation of this effect is that the first players feel entitled to take more. With sequential play, individuals adopt a first come-first served rule, whereas with simultaneous play people may adopt an equality rule. Another strategic factor is the ability to build up reputations. Research found that people take less from the common pool in public situations than in anonymous private situations. Moreover, those who harvest less gain greater prestige and influence within their group.

Structural Factors

Much research has focused on when and why people would like to structurally rearrange the commons to prevent a tragedy. Hardin stated in his analysis of the tragedy of the commons that "Freedom in a commons brings ruin to all." One of the proposed solutions is to appoint a leader to regulate access to the common. Groups are more likely to endorse a leader when a common resource is being depleted and when managing a common resource is perceived as a difficult task. Groups prefer leaders who are elected, democratic, and prototypical of the group, and these leader types are more successful in enforcing cooperation. A general aversion to autocratic leadership exists, although it may be an effective solution, possibly because of the fear of power abuse and corruption.

The provision of rewards and punishments may also be effective in preserving common resources. Selective punishments for overuse can be effective in promoting domestic water and energy conservation – for example, through installing water and electricity meters in houses. Selective rewards work, provided that they are open to everyone. An experimental carpool lane in the Netherlands failed because car commuters did not feel they were able to organize a carpool. The rewards do not have to be tangible. In Canada, utilities considered putting "smiley faces" on electricity bills of customers below the average consumption of that customer`s neighborhood.

Solutions

Articulating solutions to the tragedy of the commons is one of the main problems of political philosophy. In many situations, locals implement (often complex) social schemes that work well. The best governmental solution may be to do nothing. When these fail, there are many possible governmental solutions such as privatization, internalizing the externalities, and regulation.

Non-Governmental Solution

Sometimes the best governmental solution may be to do nothing. Robert Axelrod contends that even self-interested individuals will often find ways to cooperate, because collective restraint serves both the collective and individual interests. Anthropologist G. N. Appell criticized those who cited Hardin to "impos[e] their own economic and environmental rationality on other social systems of which they have incomplete understanding and knowledge."

Political scientist Elinor Ostrom, who was awarded 2009's Nobel Prize of Economics for her work on the issue, and others revisited Hardin's work in 1999. They found the tragedy of the commons not as prevalent or as difficult to solve as Hardin maintained, since locals have often come up with solutions to the commons problem themselves. For example, it was found that a commons in the Swiss Alps has been run by a collective of farmers there to their mutual and individual benefit since 1517, in spite of the farmers also having access to their own farmland. In general, it is in the users of a commons interests to keep the common running and complex social schemes are often invented by the users for maintaining them at optimum efficiency.

Similarly, Geographer Douglas L. Johnson remarks that many nomadic pastoralist societies of Africa and the Middle East in fact "balanced local stocking ratios against seasonal rangeland conditions in ways that were ecologically sound", reflecting a desire for lower risk rather than higher profit; in spite of this, it was often the case that "the nomad was blamed for problems that were not of his own making and were a product of alien forces." Independently finding precedent in the opinions of previous scholars such as Ibn Khaldun as well as common currency in antagonistic cultural attitudes towards non-sedentary peoples, governments and international organizations have made use of Hardin's work to help justify restrictions on land access and the eventual sedentarization of pastoral nomads despite its weak empirical basis. Examining relations between historically nomadic Bedouin Arabs and the Syrian state in the 20th century, Dawn Chatty notes that "Hardin's argument [...] was curiously accepted as the fundamental explanation for the degradation of the steppe land" in development schemes for the arid interior of the country, downplaying the larger role of agricultural overexploitation in desertification as it melded with prevailing nationalist ideology which viewed nomads as socially backward and economically harmful.

Elinor Ostrom, and her colleagues looked at how real-world communities manage communal resources, such as fisheries, land irrigation systems, and farmlands, and they identified a number of factors conducive to successful resource management. One factor is the resource itself; resources with definable boundaries (e.g., land) can be preserved much more easily. A second factor is resource dependence; there must be a perceptible threat of resource depletion, and it must be difficult to find substitutes. The third is the presence of a community; small and stable populations with a thick social network and social norms promoting conservation do better. A final condition is that there be appropriate community-based rules and procedures in place with built-in incentives for responsible use and punishments for overuse. When the commons is taken over by non-locals, those solutions can no longer be used.

Governmental Solutions

Governmental solutions may be necessary when the above conditions are not meant (such as a community being too big or too unstable to provide a thick social network). Examples of government regulation include privatization, internalizing the externalities, and regulation.

Privatization

One solution for some resources is to convert common good into private property, giving the new owner an incentive to enforce its sustainability. Libertarians and classical liberals cite the tragedy of the commons as an example of what happens when Lockean property rights to homestead resources are prohibited by a government. They argue that the solution to the tragedy of the commons is to allow individuals to take over the property rights of a resource, that is, privatizing it.

Regulation

In a typical example, governmental regulations can limit the amount of a common good that is available for use by any individual. Permit systems for extractive economic activities including mining, fishing, hunting, livestock raising and timber extraction are examples of this approach. Similarly, limits to pollution are examples of governmental intervention on behalf of the commons. This idea is used by the United Nations Moon Treaty, Outer Space Treaty and Law of the Sea

Treaty as well as the UNESCO World Heritage Convention involves the international law principle that designates some areas or resources the Common Heritage of Mankind.

In Hardin's essay, he proposed that the solution to the problem of overpopulation must be based on "mutual coercion, mutually agreed upon" and result in "relinquishing the freedom to breed". Hardin discussed this topic further in a 1979 book, *Managing the Commons,* co-written with John A. Baden. He framed this prescription in terms of needing to restrict the "reproductive right", to safeguard all other rights. Several countries have a variety of population control laws in place.

German historian Joachim Radkau thought Hardin advocates strict management of common goods via increased government involvement or international regulation bodies. An asserted impending "tragedy of the commons" is frequently warned of as a consequence for adopting policies which restrict private property and espouse expansion of public property.

Internalizing the Externalities

Privatization works when the person who owns the property (or rights of access to that property) pays the full price of its exploitation. As discussed above negative externalities (negative results, such as air or water pollution, that do not proportionately affect the user of the resource) is often a feature driving the tragedy of the commons. *Internalizing the externalities*, in other words ensuring that the users of resource pay for all of the consequences of its use, can provide an alternate solution between privatization and regulation. One example is gasoline taxes which include both the cost of road maintenance and of air pollution. This solution can provide the flexibility of privatization while minimizing the amount of government oversight and overhead that is needed.

Comedy of the Commons

In certain cases, exploiting a resource more may be a good thing. Carol M. Rose, in an 1986 article, discussed the concept of the "comedy of the commons", where the public property in question exhibits "increasing returns to scale" in usage ("the more the merrier", hence the term), in that the more people use the resource, the higher the benefit to each one. Rose cites as examples commerce and group recreational activities. According to Rose, public resources with the "comedic" characteristic may suffer from under-investment rather than over usage.

Criticism

The environmentalist Derrick Jensen claims the tragedy of the commons is used as propaganda for private ownership. He says it has been used by the political right wing to hasten the final enclosure of the "common resources" of third world and native indigenous people worldwide, as a part of the Washington Consensus. He argues that in true situations, those who abuse the commons would have been warned to desist and if they failed would have punitive sanctions against them. He says that rather than being called "The Tragedy of the Commons", it should be called "the Tragedy of the Failure of the Commons".

Hardin's work was also criticised as historically inaccurate in failing to account for the demographic transition, and for failing to distinguish between common property and open access resources. In a similar vein, Carl Dahlman argues that commons were effectively managed to prevent overgrazing. Likewise, Susan Jane Buck Cox argues that the common land example used to argue this economic concept is on very weak historical ground, and misrepresents what she terms was actually the "triumph of the commons"; the successful common usage of land for many centuries. She argues that social changes and agricultural innovation led to the demise of the commons; not the behaviour of the commoners.

Some authors, like Yochai Benkler, say with the rise of the Internet and digitalisation, make an economics system based on commons possible again. He wrote in his book *The Wealth of Networks* in 2006 that cheap computing power plus networks enable people to produce valuable products through non-commercial processes of interaction: "as human beings and as social beings, rather than as market actors through the price system". He uses the term 'networked information economy' to describe a "system of production, distribution, and consumption of information goods characterized by decentralized individual action carried out through widely distributed, nonmarket means that do not depend on market strategies." He also coined the term 'commons-based peer production' to describe collaborative efforts based on sharing information. Examples of commons-based peer production are free and open source software and open-source hardware.

Ecosystem Services

Honey bee on Avocado crop. Pollination is just one type of ecosystem service

Humankind benefits in a multitude of ways from ecosystems. Collectively, these benefits are becoming known as ecosystem services. Ecosystem services are regularly involved in the provisioning of clean drinking water and the decomposition of wastes. While scientists and environmentalists have discussed ecosystem services implicitly for decades, the ecosystem services concept itself was popularized by the Millennium Ecosystem Assessment (MA) in the early 2000s. This grouped ecosystem services into four broad categories: *provisioning*, such as the production of food and water; *regulating*, such as the control of climate and disease; *supporting*, such as nutrient cycles and crop pollination; and *cultural*, such as spiritual and recreational benefits. To help inform decision-makers, many ecosystem services are being assigned economic values.

History

While the notion of human dependence on Earth's ecosystems reaches to the start of homo sapiens' existence, the term 'natural capital' was first coined by E.F. Schumacher in 1973 in his book *Small is Beautiful* . Recognition of how ecosystems could provide more complex services to mankind date back to at least Plato (c. 400 BC) who understood that deforestation could lead to soil erosion and the drying of springs. Modern ideas of ecosystem services probably began with Marsh in 1864 when he challenged the idea that Earth's natural resources are unbounded by pointing out changes in soil fertility in the Mediterranean. It was not until the late 1940s that three key authors – Henry Fairfield Osborn, Jr, William Vogt, and Aldo Leopold – promoted recognition of human dependence on the environment.

In 1956, Paul Sears drew attention to the critical role of the ecosystem in processing wastes and recycling nutrients. In 1970, Paul Ehrlich and Rosa Weigert called attention to "ecological systems" in their environmental science textbook and "the most subtle and dangerous threat to man's existence... the potential destruction, by man's own activities, of those ecological systems upon which the very existence of the human species depends".

The term "environmental services" was introduced in a 1970 report of the *Study of Critical Environmental Problems*, which listed services including insect pollination, fisheries, climate regulation and flood control. In following years, variations of the term were used, but eventually 'ecosystem services' became the standard in scientific literature.

The ecosystem services concept has continued to expand and includes socio-economic and conservation objectives, which are discussed below. A history of the concepts and terminology of ecosystem services as of 1997, can be found in Daily's book "Nature's Services: Societal Dependence on Natural Ecosystems".

Definition

Per the 2006 Millennium Ecosystem Assessment (MA), ecosystem services are "the benefits people obtain from ecosystems." The MA also delineated the four categories of ecosystem services—supporting, provisioning, regulating and cultural—discussed below.

By 2010, there had evolved various working definitions and descriptions of ecosystem services in the literature. To prevent double counting in ecosystem services audits, for instance, The Economics of Ecosystems and Biodiversity (TEEB) replaced "Supporting Services" in the MA with "Habitat Services" and "ecosystem functions," defined as "a subset of the interactions between ecosystem structure and processes that underpin the capacity of an ecosystem to provide goods and services."

Four Categories

Detritivores like this dung beetle help to turn animal wastes into organic material that can be reused by primary producers.

The Millennium Ecosystem Assessment (MA) report 2005 defines *Ecosystem services* as benefits people obtain from ecosystems and distinguishes four categories of ecosystem services, where the so-called supporting services are regarded as the basis for the services of the other three categories. The following lists represent the definition and samples of each according to the MA:

Supporting Services

Ecosystem services "that are necessary for the production of all other ecosystem services". These include services such as nutrient recycling, primary production and soil formation. These services make it possible for the ecosystems to provide services such as food supply, flood regulation and water purification.

Provisioning Services

"Products obtained from ecosystems"

- food (including seafood and game), crops, wild foods, and spices
- raw materials (including lumber, skins, fuel wood, organic matter, fodder, and fertilizer)
- genetic resources (including crop improvement genes, and health care)
- water
- minerals (including diatomite)
- medicinal resources (including pharmaceuticals, chemical models, and test and assay organisms)
- energy (hydropower, biomass fuels)
- ornamental resources (including fashion, handicraft, jewelry, pets, worship, decoration and souvenirs like furs, feathers, ivory, orchids, butterflies, aquarium fish, shells, etc.)

Regulating Services

"Benefits obtained from the regulation of ecosystem processes"

- carbon sequestration and climate regulation

- waste decomposition and detoxification

- purification of water and air

- pest and disease control

Cultural Services

"Nonmaterial benefits people obtain from ecosystems through spiritual enrichment, cognitive development, reflection, recreation, and aesthetic experiences"

- cultural (including use of nature as motif in books, film, painting, folklore, national symbols, architect, advertising, etc.)

- spiritual and historical (including use of nature for religious or heritage value or natural)

- recreational experiences (including ecotourism, outdoor sports, and recreation)

- science and education (including use of natural systems for school excursions, and scientific discovery)

There is discussion as to how the concept of cultural ecosystem services can be operationalized. A good review of approaches in landscape aesthetics, cultural heritage, outdoor recreation, and spiritual significance to define and assess cultural values of our environment so that they fit into the ecosystem services approach is given by Daniel et al. who vote for models that explicitly link ecological structures and functions with cultural values and benefits. There also is a fundamental critique of the concept of cultural ecosystem services that builds on three arguments:

1. Pivotal cultural values attaching to the natural/cultivated environment rely on an area's unique character that cannot be addressed by methods that use universal scientific parameters to determine ecological structures and functions.

2. If a natural/cultivated environment has symbolic meanings and cultural values the object of these values are not ecosystems but shaped lifeworldly phenomena like mountains, lakes, forests, and, mainly, symbolic landscapes.

3. Those cultural values do result not from properties produced by ecosystems but are the product of a specific way of seeing within the given cultural framework of symbolic experience.

Examples

The following examples illustrate the relationships between humans and natural ecosystems through the services derived from them:

- In New York City, where the quality of drinking water had fallen below standards required by the U.S. Environmental Protection Agency (EPA), authorities opted to restore the polluted Catskill Watershed that had previously provided the city with the ecosystem service of water purification. Once the input of sewage and pesticides to the watershed area was reduced, natural abiotic processes such as soil absorption and filtration of chemicals, together with biotic recycling via root systems and soil microorganisms, water quality improved to levels that met government standards. The cost of this investment in natural capital was estimated between $1–1.5 billion, which contrasted dramatically with the estimated $6–8 billion cost of constructing a water filtration plant plus the $300 million annual running costs.

- Pollination of crops by bees is required for 15-30% of U.S. food production; most large-scale farmers import non-native honey bees to provide this service. One study reports that in California's agricultural region, it was found that wild bees alone could provide partial or complete pollination services or enhance the services provided by honey bees through behavioral interactions. However, intensified agricultural practices can quickly erode pollination services through the loss of species and those remaining are unable to compensate for the difference. The results of this study also indicate that the proportion of chaparral and oak-woodland habitat available for wild bees within 1–2 km of a farm can strongly stabilize and enhance the provision of pollination services, thereby providing a potential insurance policy for farmers of this region.

- In watersheds of the Yangtze River (China), spatial models for water flow through different forest habitats were created to determine potential contributions for hydroelectric power in the region. By quantifying the relative value of ecological parameters (vegetation-soil-slope complexes), researchers were able to estimate the annual economic benefit of maintaining forests in the watershed for power services to be 2.2 times that if it were harvested once for timber.

- In the 1980s, mineral water company Vittel (now a brand of Nestlé Waters) faced a critical problem. Nitrates and pesticides were entering the company's springs in northeastern France. Local farmers had intensified agricultural practices and cleared native vegetation that previously had filtered water before it seeped into the aquifer used by Vittel. This contamination threatened the company's right to use the "natural mineral water" label under French law. In response to this business risk, Vittel developed an incentive package for farmers to improve their agricultural practices and consequently reduce water pollution that had affected Vittel's product. For example, Vittel provided subsidies and free technical assistance to farmers in exchange for farmers' agreement to enhance pasture management, reforest catchments, and reduce the use of agrochemicals. This is an example of a Payment for ecosystem services program.

Ecology

Understanding of ecosystem services requires a strong foundation in ecology, which describes the underlying principles and interactions of organisms and the environment. Since the scales at which these entities interact can vary from microbes to landscapes, milliseconds to millions of years, one of the greatest remaining challenges is the descriptive characterization of energy and material flow between them. For example, the area of a forest floor, the detritus upon it, the microorganisms in the soil and characteristics of the soil itself will all contribute to the abilities of that forest for providing ecosystem services like carbon sequestration, water purification, and erosion prevention to other areas within the watershed. Note that it is often possible for multiple services to be bundled together and when benefits of targeted objectives are secured, there may also be ancillary benefits – the same forest may provide habitat for other organisms as well as human recreation, which are also ecosystem services.

The complexity of Earth's ecosystems poses a challenge for scientists as they try to understand how relationships are interwoven among organisms, processes and their surroundings. As it relates to human ecology, a suggested research agenda for the study of ecosystem services includes the following steps:

1. identification of *ecosystem service providers* (*ESPs*) – species or populations that provide specific ecosystem services – and characterization of their functional roles and relationships;

2. determination of community structure aspects that influence how ESPs function in their natural landscape, such as compensatory responses that stabilize function and non-random extinction sequences which can erode it;

3. assessment of key environmental (abiotic) factors influencing the provision of services;

4. measurement of the spatial and temporal scales ESPs and their services operate on.

Recently, a technique has been developed to improve and standardize the evaluation of ESP functionality by quantifying the relative importance of different species in terms of their efficiency and abundance. Such parameters provide indications of how species respond to changes in the environment (i.e. predators, resource availability, climate) and are useful for identifying species that are disproportionately important at providing ecosystem services. However, a critical drawback is that the technique does not account for the effects of interactions, which are often both complex and fundamental in maintaining an ecosystem and can involve species that are not readily detected as a priority. Even so, estimating the functional structure of an ecosystem and combining it with information about individual species traits can help us understand the resilience of an ecosystem amidst environmental change.

Many ecologists also believe that the provision of ecosystem services can be stabilized with biodiversity. Increasing biodiversity also benefits the variety of ecosystem services available to society. Understanding the relationship between biodiversity and an ecosystem's stability is essential to the management of natural resources and their services.

Redundancy Hypothesis

The concept of ecological redundancy is sometimes referred to as *functional compensation* and assumes that more than one species performs a given role within an ecosystem. More specifically, it is characterized by a particular species increasing its efficiency at providing a service when conditions are stressed in order to maintain aggregate stability in the ecosystem. However, such increased dependence on a compensating species places additional stress on the ecosystem and often enhances its susceptibility to subsequent disturbance. The redundancy hypothesis can be summarized as "species redundancy enhances ecosystem resilience".

Another idea uses the analogy of rivets in an airplane wing to compare the exponential effect the loss of each species will have on the function of an ecosystem; this is sometimes referred to as *rivet popping*. If only one species disappears, the loss of the ecosystem's efficiency as a whole is relatively small; however if several species are lost, the system essentially collapses as an airplane wing would, were it to lose too many rivets. The hypothesis assumes that species are relatively specialized in their roles and that their ability to compensate for one another is less than in the redundancy hypothesis. As a result, the loss of any species is critical to the performance of the ecosystem. The key difference is the rate at which the loss of species affects total ecosystem function.

Portfolio Effect

A third explanation, known as the *portfolio effect*, compares biodiversity to stock holdings, where diversification minimizes the volatility of the investment, or in this case, the risk in stability of ecosystem services. This is related to the idea of *response diversity* where a suite of species will exhibit differential responses to a given environmental perturbation and therefore when considered together, they create a stabilizing function that preserves the integrity of a service.

Several experiments have tested these hypotheses in both the field and the lab. In ECOTRON, a laboratory in the UK where many of the biotic and abiotic factors of nature can be simulated, studies have focused on the effects of earthworms and symbiotic bacteria on plant roots. These laboratory experiments seem to favor the rivet hypothesis. However, a study on grasslands at Cedar Creek Reserve in Minnesota seems to support the redundancy hypothesis, as have many other field studies.

Economics

Sustainable urban drainage pond near housing in Scotland. The filtering and cleaning of surface and waste water by natural vegetation is a form of ecosystem service.

There are questions regarding the environmental and economic values of ecosystem services. Some people may be unaware of the environment in general and humanity's interrelatedness with the natural environment, which may cause misconceptions. Although environmental awareness is rapidly improving in our contemporary world, ecosystem capital and its flow are still poorly understood, threats continue to impose, and we suffer from the so-called 'tragedy of the commons'. Many efforts to inform decision-makers of current versus future costs and benefits now involve organizing and translating scientific knowledge to economics, which articulate the consequences of our choices in comparable units of impact on human well-being. An especially challenging aspect of this process is that interpreting ecological information collected from one spatial-temporal scale does not necessarily mean it can be applied at another; understanding the dynamics of ecological processes relative to ecosystem services is essential in aiding economic decisions. Weighting factors such as a service's irreplaceability or bundled services can also allocate economic value such that goal attainment becomes more efficient.

The economic valuation of ecosystem services also involves social communication and information, areas that remain particularly challenging and are the focus of many researchers. In general, the idea is that although individuals make decisions for any variety of reasons, trends reveal the aggregative preferences of a society, from which the economic value of services can be inferred and assigned. The six major methods for valuing ecosystem services in monetary terms are:

- Avoided cost: Services allow society to avoid costs that would have been incurred in the absence of those services (e.g. waste treatment by wetland habitats avoids health costs)

- Replacement cost: Services could be replaced with man-made systems (e.g. restoration of the Catskill Watershed cost less than the construction of a water purification plant)

- Factor income: Services provide for the enhancement of incomes (e.g. improved water quality increases the commercial take of a fishery and improves the income of fishers)

- Travel cost: Service demand may require travel, whose costs can reflect the implied value of the service (e.g. value of ecotourism experience is at least what a visitor is willing to pay to get there)

- Hedonic pricing: Service demand may be reflected in the prices people will pay for associated goods (e.g. coastal housing prices exceed that of inland homes)

- Contingent valuation: Service demand may be elicited by posing hypothetical scenarios that involve some valuation of alternatives (e.g. visitors willing to pay for increased access to national parks)

A peer-reviewed study published in 1997 estimated the value of the world's ecosystem services and natural capital to be between US$16–54 trillion per year, with an average of US$33 trillion per year. However, Salles (2011) indicates 'The total value of biodiversity is infinite, so having debate about what is the total value of nature is actually pointless because we can't live without it'.

Management and Policy

Although monetary pricing continues with respect to the valuation of ecosystem services, the challenges in policy implementation and management are significant and multitudinous. The administration of common pool resources is a subject of extensive academic pursuit. From defining the problems to finding solutions that can be applied in practical and sustainable ways, there is much to overcome. Considering options must balance present and future human needs, and decision-makers must frequently work from valid but incomplete information. Existing legal policies are often considered insufficient since they typically pertain to human health-based standards that are mismatched with necessary means to protect ecosystem health and services. To improve the information available, one suggestion has involved the implementation of an *Ecosystem Services Framework* (ESF), which integrates the biophysical and socio-economic dimensions of protecting the environment and is designed to guide institutions through multidisciplinary information and jargon, helping to direct strategic choices.

Novel and expedient methods are needed to deal with managing Earth's ecosystem services. Local to regional collective management efforts might be considered appropriate for services like crop pollination or resources like water. Another approach that has become increasingly popular over the last decade is the marketing of ecosystem services protection. Payment and trading of services is an emerging worldwide small-scale solution where one can acquire credits for activities such as sponsoring the protection of carbon sequestration sources or the restoration of ecosystem service providers. In some cases, banks for handling such credits have been established and conservation companies have even gone public on stock exchanges, defining an evermore parallel link with economic endeavors and opportunities for tying into social perceptions. However, crucial for implementation are clearly defined land rights, which is often lacking in many developing countries. In particular, many forest-rich developing countries suffering deforestation experience conflict between different forest stakeholders. In addition, concerns for such global transactions include inconsistent compensation for services or resources sacrificed elsewhere and misconceived warrants for irresponsible use. Another approach has been focused on protecting ecosystem service 'hotspots'. Recognition that the conservation of many ecosystem services aligns with more traditional conservation goals (i.e. biodiversity) has led to the suggested merging of objectives for maximizing their mutual success. This may be particularly strategic when employing networks that permit the flow of services across landscapes, and might also facilitate securing the financial means to protect services through a diversification of investors.

For example, in recent years there has been interest in the valuation of ecosystem services provided by shellfish production and restoration. A keystone species, low in the food chain, bivalve shellfish such as oysters support a complex community of species by performing a number of functions essential to the diverse array of species that surround them. There is also increasing recognition that some shellfish species may impact or control many ecological processes; so much so that they are included on the list of "ecosystem engineers"—organisms that physically, biologically or chemically modify the environment around them in ways that influence the health of other organisms. Many of the ecological functions and processes performed or affected by shellfish contribute to human well-being by providing a stream of valuable ecosystem services over time by filtering out particulate materials and potentially mitigating water quality issues by controlling excess nutrients in the water.

Ecosystem-Based Adaptation (EBA)

Ecosystem-Based Adaptation or EbA is an emerging strategy for community development and environmental management that seeks to use an ecosystem services framework to help communities adapt to the effects of climate change. The Convention on Biological Diversity currently defines Ecosystem-Based Adaptation as "the use of biodiversity and ecosystem services to help people adapt to the adverse effects of climate change", which includes the use of "sustainable management, conservation and restoration of ecosystems, as part of an overall adaptation strategy that takes into account the multiple social, economic and cultural co-benefits for local communities".

In 2001, the Millennium Ecosystem Assessment announced that humanity's impact on the natural world was increasing to levels never before seen, and that the degradation of the planet's ecosystems would become a major barrier to achieving the Millennium Development Goals. In recognition of this fact, Ecosystem-Based Adaptation seeks to use the restoration of ecosystems as a stepping-stone to improving the quality of life in communities experiencing the impacts of climate change. Specifically, this involves the restoration of ecosystems that provide the community with essential services, such as the provisioning of food and water and protection from storm surges and flooding. EbA interventions typically combine elements of both climate change mitigation and adaptation to global warming to help address the community's current and future needs.

Collaborative planning between scientists, policy makers, and community members is an essential element of Ecosystem-Based Adaptation. By drawing on the expertise of outside experts and local residents alike, EbA seeks to develop unique solutions to unique problems, rather than simply replicating past projects.

Estuarine and Coastal Ecosystem Services

Ecosystem services are defined as the gains acquired by humankind from surroundings ecosystems. Four different types of ecosystem services have been distinguished by the scientific body: regulating services, provisioning services, cultural services and supporting services. An ecosystem does not necessarily offer all four types of services simultaneously; but given the intricate nature of any ecosystem, it is usually assumed that humans benefit from a combination of these services. The services offered by diverse types of ecosystems (forests, seas, coral reefs, mangroves, etc.) differ in nature and in consequence. In fact, some services directly affect the livelihood of neighboring human populations (such as fresh water, food or aesthetic value, etc.) while other services affect general environmental conditions by which humans are indirectly impacted (such as climate change, erosion regulation or natural hazard regulation, etc.).

Estuarine and coastal ecosystems are both marine ecosystems. An estuary is defined as the area in which a river meets the sea or the ocean. The waters surrounding this area are predominantly salty waters or brackish waters; and the incoming river water is dynamically motioned by the tide. An estuary strip may be covered by populations of reed (or similar plants) and/or sandbanks (or similar form or land).

A coastal ecosystem occurs in areas where the sea or ocean waters meet the land.

Regulating Services

Regulating services are the "benefits obtained from the regulation of ecosystem processes". In the case of coastal and estuarine ecosystems, these services include climate regulation, waste treatment and disease control and natural hazard regulation.

- Climate Regulation

Both the biotic and abiotic ensembles of marine ecosystems play a role in climate regulation. They act as sponges when it comes to gases in the atmosphere, retaining large levels of CO_2 and other Green House Gases (methane and nitrous oxide). Marine plants also use CO_2 for photosynthesis purposes and help in reducing the atmospheric CO_2. The oceans and seas absorb the heat from the atmosphere and redistribute it through the means of water currents, and atmospheric processes, such as evaporation and the reflection of light allow for the cooling and warming of the overlying atmosphere. The ocean temperatures are thus imperative to the regulation of the atmospheric temperatures in any part of the world: "without the ocean, the Earth would be unbearably hot during the daylight hours and frigidly cold, if not frozen, at night".

- Waste Treatment & Disease Regulation

Another service offered by marine ecosystem is the treatment of wastes, thus helping in the regulation of diseases. Wastes can be diluted and detoxified through transport across marine ecosystems; pollutants are removed from the environment and stored, buried or recycled in marine ecosystems: "Marine ecosystems break down organic waste through microbial communities that filter water, reduce/limit the effects of eutrophication, and break down toxic hydrocarbons into their basic components such as carbon dioxide, nitrogen, phosphorus, and water". The fact that waste is diluted with large volumes of water and moves with water currents leads to the regulation of diseases and the reduction of toxics in seafood.

- Buffer Zones

Coastal and estuarine ecosystems act as buffer zones against natural hazards and environmental disturbances, such as floods, cyclones, tidal surges and storms. The role they play is to "[absorb] a portion of the impact and thus [lessen] its effect on the land". Wetlands, for example, and the vegetation it supports – trees, root mats, etc. – retain large amounts of water (surface water, snowmelt, rain, groundwater) and then slowly releases them back, decreasing the likeliness of floods. Mangrove forests protect coastal shorelines from tidal erosion or erosion by currents; a process that was studied after the 1999 cyclone that hit India. Villages that were surrounded with mangrove forests encountered less damages than other villages that weren't protected by mangroves.

Provisioning Services

Provisioning services consist of all "the products obtained from ecosystems". Marine ecosystems provide people with: wild & cultured seafood, fresh water, fiber & fuel and biochemical & genetic resources.

- Marine Products

Humans consume a large number of products originating from the seas, whether as a nutritious product or for use in other sectors: "More than one billion people worldwide, or one-sixth of the global population, rely on fish as their main source of animal protein. In 2000, marine and coastal fisheries accounted for 12 per cent of world food production". Fish and other edible marine products - primarily fish, shellfish, roe and seaweeds – constitute for populations living along the coast the main elements of the local cultural diets, norms and traditions. A very pertinent example would be Sushi, the national food of Japan, which consists mostly of different types of fish and seaweed.

- Fresh Water

Water bodies that are not highly concentrated in salts are referred to as 'fresh water' bodies. Fresh water may run through lakes, rivers and streams, to name a few; but it is most prominently found in the frozen state or as soil moisture or buried deep underground. Fresh water is not only important for the survival of humans, but also for the survival of all the existing species of animals, plants.

- Raw Materials

Marine creatures provide us with the raw materials needed for the manufacturing of clothing, building materials (lime extracted from coral reefs), ornamental items and personal-use items (luffas, art and jewelry): "The skin of marine mammals for clothing, gas deposits for energy production, lime (extracted from coral reefs) for building construction, and the timber of mangroves and coastal forests for shelter are some of the more familiar uses of marine organisms. Raw marine materials are utilized for non-essential goods as well, such as shells and corals in ornamental items". Humans have also referred to processes within marine environments for the production of renewable energy: using the power of waves – or tidal power – as a source of energy for the powering of a turbine, for example. Oceans and seas are used as sites for offshore oil and gas installations, offshore wind farms.

- Biochemical and Genetic Resources

Biochemical resources are compounds extracted from marine organisms for use in medicines, pharmaceuticals, cosmetics and other biochemical products. Genetic resources are the genetic information found in marine organisms that would later on be used for animal and plant breeding and for technological advances in the biological field. These resources are either directly taken out from an organism – such as fish oil as a source of omega3 –, or used as a model for innovative man-made products: "such as the construction of fiber optics technology based on the properties of sponges. Compared to terrestrial products, marine-sourced products tend to be more highly bioactive, likely due to the fact that marine organisms have to retain their potency despite being diluted in the surrounding sea-water".

Cultural Services

Cultural services relate to the non-material world, as they benefit the benefit recreational, aesthetic, cognitive and spiritual activities, which are not easily quantifiable in monetary terms.

- Inspirational

Marine environments have been used by many as an inspiration for their works of art, music, architecture, traditions... Water environments are spiritually important as a lot of people view them as a means for rejuvenation and change of perspective. Many also consider the water as being a part of their personality, especially if they have lived near it since they were kids: they associate it to fond memories and past experiences. Living near water bodies for a long time results in a certain set of water activities that become a ritual in the lives of people and of the culture in the region.

- Recreation and Tourism

Sea sports are very popular among coastal populations: surfing, snorkeling, whale watching, kayaking, recreational fishing...a lot of tourists also travel to resorts close to the sea or rivers or lakes to be able to experience these activities, and relax near the water.

Beach accommodated into a recreational area.

- Science and Education

A lot can be learned from marine processes, environments and organisms – that could be implemented into our daily actions and into the scientific domain. Although much is still yet to still be known about the ocean world: "by the extraordinary intricacy and complexity of the marine environment and how it is influenced by large spatial scales, time lags, and cumulative effects".

Supporting Services

Supporting services are the services that allow for the other ecosystem services to be present. They have indirect impacts on humans that last over a long period of time. Several services can be considered as being both supporting services and regulating/cultural/provisioning services.

- Nutrient Cycling

"Nutrient cycling refers to the storage, cycling, and maintenance of nutrients by organisms and their associated processes". The ocean is a vast storage pool for these nutrients, such as carbon, nitrogen and phosphorus. The nutrients are absorbed by the basic organisms of the marine food web and are thus transferred from one organism to the other and from one ecosystem to the other. Nutrients are recycled through the life cycle of organisms as they die and decompose, releasing the nutrients into the neighboring environment. "The service of nutrient cycling eventually impacts all other ecosystem services as all living things require a constant supply of nutrients to survive".

- Biologically Mediated Habitats

Biologically mediated habitats are defined as being the habitats that living marine structures offer to other organisms. These need not to be designed for the sole purpose of serving as a habitat, but happen to become living quarters whilst growing naturally. For example, coral reefs and mangrove forests are home to numerous species of fish, seaweed and shellfish... The importance of these habitats is that they allow for interactions between different species, aiding the provisioning of marine goods and services. They are also very important for the growth at the early life stages of marine species (breeding and bursary spaces), as they serve as a food source and as a shelter from predators.

Coral and other living organisms serve as habitats for many marine species.

- Primary Production

Primary production refers to the production of organic matter, i.e., chemically bound energy, through processes such as photosynthesis and chemosynthesis. The organic matter produced by primary producers forms the basis of all food webs. Further, it generates oxygen, a molecule necessary for life.

Businessworld

Ecosystem services degradation can pose a number of risks to corporate performance as well as provide business opportunities through ecosystem restoration and enhancement. Risks and opportunities include:

- **Operational**
  - Risks such as higher costs for freshwater due to scarcity or lower output for hydro-electric facilities due to siltation
  - Opportunities such as increasing water-use efficiency or building an on-site wetland to circumvent the need for new water treatment infrastructure

- **Regulatory and Legal**
  - Risks such as new fines, government regulations, or lawsuits from local communities that lose ecosystem services due to corporate activities
  - Opportunities such as engaging governments to develop policies and incentives to protect or restore ecosystems that provide services a company needs

- **Reputational**
  - Risks such as retail companies being targeted by nongovernmental organization campaigns for purchasing wood or paper from sensitive forests

Opportunities such as implementing and communicating sustainable purchasing, operating, or investment practices in order to differentiate corporate brands.

- **Market and Product**
  - Risks such as customers switching to other suppliers that offer products with lower ecosystem impacts or governments implementing new sustainable procurement policies
  - Opportunities such as launching new products and services that reduce customer impacts on ecosystems or participating in emerging markets for carbon sequestration and watershed protection other products

- **Financing**
  - Risks such as banks implementing more rigorous lending requirements for corporate loans
  - Opportunities such as banks offering more favorable loan terms or investors taking positions in companies supplying products and services that improve resource use efficiency or restore degraded ecosystems

Many companies are not fully aware of the extent of their dependence and impact on ecosystems and the possible ramifications. Likewise, environmental management systems and environmental due diligence tools are more suited to handle "traditional" issues of pollution and natural resource consumption. Most focus on environmental impacts, not dependence. Several newly developed tools and methodologies can help the private sector value and assess ecosystem services. These include Our Ecosystem, the Corporate Ecosystem Services Review (ESR), Artificial Intelligence for Ecosystem Services (ARIES), the Natural Value Initiative (NVI) and InVEST (Integrated Valuation of Ecosystem Services & Tradeoffs)

Land Use Change Decisions

Ecosystem services decisions require making complex choices at the intersection of ecology, technology, society and the economy. The process of making ecosystem services decisions must consider the interaction of many types of information, honor all stakeholder viewpoints, including regulatory agencies, proposal proponents, decision makers, residents, NGOs, and measure the impacts on all four parts of the intersection. These decisions are usually spatial, always multi-objective, and based on uncertain data, models, and estimates. Often it is the combination of the best science combined with the stakeholder values, estimates and opinions that drive the process.

One analytical study modeled the stakeholders as agents to support water resource management decisions in the Middle Rio Grande basin of New Mexico. This study focused on modeling the stakeholder inputs across a spatial decision, but ignored uncertainty. Another study used Monte Carlo methods to exercise econometric models of landowner decisions in a study of the effects of land-use change. Here the stakeholder inputs were modeled as random effects to reflect the uncertainty. A third study used a Bayesian decision support system to both model the uncertainty in the scientific information Bayes Nets and to assist collecting and fusing the input from stakeholders. This study was about siting wave energy devices off the Oregon Coast, but presents a general method for managing uncertain spatial science and stakeholder information in a decision making environment. Remote sensing data and analyses can be used to assess the health and extent of land cover classes that provide ecosystem services, which aids in planning, management, monitoring of stakeholders' actions, and communication between stakeholders.

In Baltic countries scientists, nature conservationists and local authorities are implementing integrated planning approach for grassland ecosystems. They are develping Integrated Planning Tool that will be based on GIS (geographic information system) technology and put online that will help for planners to choose the best grassland management solution for concrete grassland. It will look holistically at the processes in the countryside and help to find best grassland management solutions by taking into account both natural and socioeconomic factors of the particular site.

Habitat Conservation

Tree planting is an aspect of habitat conservation. In each plastic tube a hardwood tree has been planted.

There are significant ecological benefits associated with selective cutting.
Pictured is an area with Ponderosa Pine trees that were selectively harvested.

Habitat conservation is a management practice that seeks to conserve, protect and restore habitat areas for wild plants and animals, especially conservation reliant species, and prevent their extinction, fragmentation or reduction in range. It is a priority of many groups that cannot be easily characterized in terms of any one ideology.

History of the Conservation Movement

For much of human history, *nature* had been seen as a resource, one that could be controlled by the government and used for personal and economic gain. The idea was that plants only existed to feed animals and animals only existed to feed humans. The land itself had limited value only extending to the resources it could provide such as minerals and oil.

Throughout the 18th and 19th centuries social views started to change and scientific conservation principles were first practically applied to the forests of British India. The conservation ethic that began to evolve included three core principles: that human activity damaged the environment, that there was a civic duty to maintain the environment for future generations, and that scientific, empirically based methods should be applied to ensure this duty was carried out. Sir James Ranald Martin was prominent in promoting this ideology, publishing many medico-topographical reports that demonstrated the scale of damage wrought through large-scale deforestation and desiccation, and lobbying extensively for the institutionalization of forest conservation activities in British India through the establishment of Forest Departments.

The Madras Board of Revenue started local conservation efforts in 1842, headed by Alexander Gibson, a professional botanist who systematically adopted a forest conservation program based on scientific principles. This was the first case of state conservation management of forests in the world. Governor-General Lord Dalhousie introduced the first permanent and large-scale forest conservation program in the world in 1855, a model that soon spread to other colonies, as well the United States, where Yellowstone National Park was opened in 1872 as the world's first national park.

Rather than focusing on the economic or material benefits associated with nature, humans began to appreciate the value of nature itself and the need to protect pristine wilderness. By the middle of the 20th century countries such as the United States, Canada, and Britain understood this appreciation and instigated laws and legislation in order to ensure that the most fragile and beautiful environments would be protected for generations to come. Today with the help of NGO's, not-for-profit organizations and governments world-wide there is a stronger movement taking place, with a deeper understanding of habitat conservation with the aim of protecting delicate habitats and preserving biodiversity on a global scale. The commitment and actions of small volunteering association in villages and towns, that endeavour to emulate the work done by well known Conservation Organisations, is paramount in ensuring generations that follow understand the importance of conserving natural resources. A village conservation group with the mission statement "We are committed to protecting and enhancing the natural environment in and around the adjoining villages of Ouston and Urpeth." may one day inspire a child who becomes the employee of a world-wide conservation organisation.

Values of Natural Habitat

The natural environment is a source for a wide range of resources that can be exploited for economic profit, for example timber is harvested from forests and clean water is obtained from natural streams. However, land development from anthropogenic economic growth often causes a decline in the ecological integrity of nearby natural habitat. For instance, this was an issue in the northern rocky mountains of the USA.

However, there is also economic value in conserving natural habitats. Financial profit can be made from tourist revenue, particularly in the tropics where species diversity is high. The cost of repairing damaged ecosystems is considered to be much higher than the cost of conserving natural ecosystems.

Measuring the worth of conserving different habitat areas is often criticized as being too utilitarian from a philosophical point of view.

Biodiversity

Habitat conservation is important in maintaining biodiversity, an essential part of global food security. There is evidence to support a trend of accelerating erosion of the genetic resources of agricultural plants and animals. An increase in genetic similarity of agricultural plants and animals means an increased risk of food loss from major epidemics. Wild species of agricultural plants have been found to be more resistant to disease, for example the wild corn species Teosinte is resistant to 4 corn diseases that affect human grown crops. A combination of seed banking and habitat conservation has been proposed to maintain plant diversity for food security purposes.

Classifying Environmental Values

Pearce and Moran outlined the following method for classifying environmental uses:

- Direct extractive uses: e.g. timber from forests, food from plants and animals

- Indirect uses: e.g. ecosystem services like flood control, pest control, erosion protection

- Optional uses: future possibilities e.g. unknown but potential use of plants in chemistry/medicine

- Non-use values:

  o Bequest value (benefit of an individual who knows that others may benefit from it in future)

  o Passive use value (sympathy for natural environment, enjoyment of the mere existence of a particular species)

Impacts

Natural Causes

Habitat loss and destruction can occur both naturally and through anthropogenic causes. Events leading to natural habitat loss include climate change, catastrophic events such as volcanic explosions and through the interactions of invasive and non-invasive species. Natural climate change, events have previously been the cause of many widespread and large scale losses in habitat. For example, some of the mass extinction events generally referred to as the "Big Five" have coincided with large scale such as the Earth entering an ice age, or alternate warming events. Other events in the big five also have their roots in natural causes, such as volcanic explosions and meteor collisions. The Chicxulub impact is one such example, which has previously caused widespread losses in habitat as the Earth either received less sunlight or grew colder, causing certain fauna and flora to flourish whilst others perished. Previously known warm areas in the tropics, the most sensitive habitats on Earth, grew colder, and areas such as Australia developed radically different flora and fauna to those seen today. The big five mass extinction events have also been linked to sea level changes, indicating that large scale marine species loss was strongly influenced by loss in marine habitats, particularly shelf habitats. Methane-driven oceanic eruptions have also been shown to have caused smaller mass extinction events.

Human Impacts

Humans have been the cause of many species' extinction. Due to humans' changing and modifying their environment, the habitat of other species often become altered or destroyed as a result of human actions. Even before the modern industrial era, humans were having widespread, and major effects on the environment. A good example of this is found in Aboriginal Australians and Australian megafauna. Aboriginal hunting practices, which included burning large sections of forest at a time, eventually altered and changed Australia's vegetation so much that many herbivorous megafauna species were left with no habitat and were driven into extinction. Once herbivorous megafauna species became extinct, carnivorous megafauna species soon followed. In the recent past, humans have been responsible for causing more extinctions within a given period of time than ever before. Deforestation, pollution, anthropogenic climate change and human settlements have all been driving forces in altering or destroying habitats. The destruction of ecosystems such as rainforests has resulted in countless habitats being destroyed. These biodiversity hotspots are home to millions of habitat specialists, which do not exist beyond a tiny area. Once their habitat is destroyed, they cease to exist.This destruction has a follow-on effect, as species which coexist or depend upon the existence of other species also become extinct, eventually resulting in the collapse of an entire ecosystem. These time-delayed extinctions are referred to as the extinction debt, which is the result of destroying and fragmenting habitats. As a result of anthropogenic modification of the environment, the extinction rate has climbed to the point where the Earth is now within a sixth mass extinction event, as commonly agreed by biologists. This has been particularly evident, for example, in the rapid decline in the number of amphibian species worldwide.

Approaches and Methods of Habitat Conservation

Determining the size, type and location of habitat to conserve is a complex area of conservation biology. Although difficult to measure and predict, the conservation value of a habitat is often a reflection of the quality (e.g. species abundance and diversity), endangerment of encompassing ecosystems, and spatial distribution of that habitat.

Identifying Priority Habitats for Conservation

Habitat conservation is vital for protecting species and ecological processes. It is important to conserve and protect the space/ area in which that species occupies. Therefore, areas classified as 'biodiversity hotspots', or those in which a flagship, umbrella, or endangered species inhabits are often the habitats that are given precedence over others. Species that possess an elevated risk of extinction are given the highest priority and as a result of conserving their habitat, other species in that community are protected thus serving as an element of gap analysis. In the United States of America, a Habitat Conservation Plan (HCP) is often developed to conserve the environment in which a specific species inhabits. Under the U.S. Endangered Species Act (ESA) the habitat that requires protection in an HCP is referred to as the 'critical habitat'. Multiple-species HCPs are becoming more favourable than single-species HCPs as they can potentially protect an array of species before they warrant listing under the ESA, as well as being able to conserve broad ecosystem components and processes . As of January 2007, 484 HCPs were permitted across the United States, 40 of which covered 10 or more species.The San Diego Multiple Species Conservation Plan (MSCP) encompasses 85 species in a total area of 26,000-km2. Its aim is to protect the habitats of multiple species and overall biodiversity by minimizing development in sensitive areas.

HCPs require clearly defined goals and objectives, efficient monitoring programs, as well as successful communication and collaboration with stakeholders and land owners in the area. Reserve design is also important and requires a high level of planning and management in order to achieve the goals of the HCP. Successful reserve design often takes the form of a hierarchical system with the most valued habitats requiring high protection being surrounded by buffer habitats that have a lower protection status. Like HCPs, hierarchical reserve design is a method most often used to protect a single species, and as a result habitat corridors are maintained, edge effects are reduced and a broader suite of species are protected.

How Much Habitat is Needed

A range of methods and models currently exist that can be used to determine how much habitat is to be conserved in order to sustain a viable population. Modelling tools often rely on the spatial scale of the area as an indicator of conservation value. There has been an increase in emphasis on conserving few large areas of habitat as opposed to many small areas. This idea is often referred to as the "single large or several small", SLOSS debate, and is a highly controversial area among conservation biologists and ecologists. The reasons behind the argument that "larger is better" include the reduction in the negative impacts of patch edge effects, the general idea that species richness increases with habitat area and the ability of larger habitats to support greater populations with lower extinction probabilities. Noss & Cooperrider support the "larger is better" claim and developed a model that implies areas of habitat less than 1000ha are "tiny" and of low conservation value. However, Shwartz suggests that although "larger is better", this does not imply that "small is bad". Shwartz argues that human induced habitat loss leaves no alternative to conserving small areas. Furthermore, he suggests many endangered species which are of high conservation value, may only be restricted to small isolated patches of habitat, and thus would be overlooked if larger areas were given a higher priority. The shift to conserving larger areas is somewhat justified in society by placing more value on larger vertebrate species, which naturally have larger habitat requirements.

Examples of Current Conservation Organizations

The Nature Conservancy

Since its formation in 1951 The Nature Conservancy has slowly developed into one of the world's largest conservation organizations. Currently operating in over 30 countries, across 5 continents world-wide, The Nature Conservancy aims to protect nature and its assets for future generations. The organization purchases land or accepts land donations with the intension of conserving its natural resources. In 1955 The Nature Conservancy purchased its first 60-acre plot near the New York/Connecticut border in the United States of America. Today the Conservancy has expanded to protect over 119 million acres of land, 5,000 river miles as well as participating in over 1000 marine protection programs across the globe . Since its beginnings The Nature Conservancy has understood the benefit in taking a scientific approach towards habitat conservation. For the last decade the organization has been using a collaborative, scientific method known as 'Conservation by Design'. By collecting and analyzing scientific data The Conservancy is able to holistically approach the protection of various ecosystems. This process determines the habitats that need protection, specific elements that should be conserved as well as monitoring progress so more efficient practices can be developed for the future.

The Nature Conservancy currently has a large number of diverse projects in operation. They work with countries around the world to protect forests, river systems, oceans, deserts and grasslands. In all cases the aim is to provide a sustainable environment for both the plant and animal life forms that depend on them as well as all future generations to come. turtles

World Wildlife Fund (WWF)

The World Wildlife Fund (WWF) was first formed in after a group of passionate conservationists signed what is now referred to as the Morges Manifesto. WWF is currently operating in over 100 countries across 5 continents with a current listing of over 5 million supporters. One of the first projects of WWF was assisting in the creation of the Charles Darwin Research Foundation which aided in the protection of diverse range of unique species existing on the

Galápagos' Islands, Ecuador. It was also a WWF grant that helped with the formation of the College of African Wildlife Management in Tanzania which today focuses on teaching a wide range of protected area management skills in areas such as ecology, range management and law enforcement. The WWF has since gone on to aid in the protection of land in Spain, creating the Coto Doñana National Park in order to conserve migratory birds and The Democratic Republic of Congo, home to the world's largest protected wetlands. The WWF also initiated a debt-for-nature concept which allows the country to put funds normally allocated to paying off national debt, into conservation programs that protect its natural landscapes. Countries currently participating include Madagascar, the first country to participate which since 1989 has generated over $US50 million towards preservation, Bolivia, Costa Rica, Ecuador, Gabon, the Philippines and Zambia.

Rare Conservation

Rare has been in operation since 1973 with current global partners in over 50 countries and offices in the United States of America, Mexico, the Philippines, China and Indonesia. Rare focuses on the human activity that threatens biodiversity and habitats such as overfishing and unsustainable agriculture. By engaging local communities and changing behaviour Rare has been able to launch campaigns to protect areas in most need of conservation. The key aspect of Rare's methodology is their "Pride Campaign's". For example, in the Andes in South America, Rare has partnered with 11 different sites with the intention of creating incentives to develop watershed protection practices. In the Southeast Asia's "coral triangle" Rare is training fishers in local communities to better manage the areas around the coral reefs in order to lessen human impact. Such programs last for three years with the aim of changing community attitudes so as to conserve fragile habitats and provide ecological protection for years to come.

WWF Netherlands

WWF Netherlands, along with ARK Nature, Wild Wonders of Europe and Conservation Capital have started the Rewilding Europe project. This project intents to rewild several areas in Europe.

Environmental Management System

Environmental management system (EMS) refers to the management of an organization's environmental programs in a comprehensive, systematic, planned and documented manner. It includes the organizational structure, planning and resources for developing, implementing and maintaining policy for environmental protection.

More formally, EMS is "a system and database which integrates procedures and processes for training of personnel, monitoring, summarizing, and reporting of specialized environmental performance information to internal and external stakeholders of a firm."

The most widely used standard on which an EMS is based is International Organization for Standardization (ISO) 14001. Alternatives include the EMAS.

An environmental management information system (EMIS) is an information technology solution for tracking environmental data for a company as part of their overall environmental management system.

Goals

The goals of EMS are to increase compliance and reduce waste:

- Compliance is the act of reaching and maintaining minimal legal standards. By not being compliant, companies may face fines, government intervention or may not be able to operate.

- Waste reduction goes beyond compliance to reduce environmental impact. The EMS helps to develop, implement, manage, coordinate and monitor environmental policies. Waste reduction begins at the design phase through pollution prevention and waste minimization. At the end of the life cycle, waste is reduced by recycling.

Features

An environmental management system (EMS):

- Serves as a tool, or process, to improve environmental performance and information mainly "design, pollution control and waste minimization, training, reporting to top management, and the setting of goals"

- Provides a systematic way of managing an organization's environmental affairs

- Is the aspect of the organization's overall management structure that addresses immediate and long-term impacts of its products, services and processes on the environment. EMS assists with planning, controlling and monitoring policies in an organization.

- Gives order and consistency for organizations to address environmental concerns through the allocation of resources, assignment of responsibility and ongoing evaluation of practices, procedures and processes

- Creates environmental buy-in from management and employees and assigns accountability and responsibility.

- Sets framework for training to achieve objectives and desired performance.

- Helps understand legislative requirements to better determine a product or service's impact, significance, priorities and objectives.

- Focuses on continual improvement of the system and a way to implement policies and objectives to meet a desired result. This also helps with reviewing and auditing the EMS to find future opportunities.

- Encourages contractors and suppliers to establish their own EMS.

EMS Model

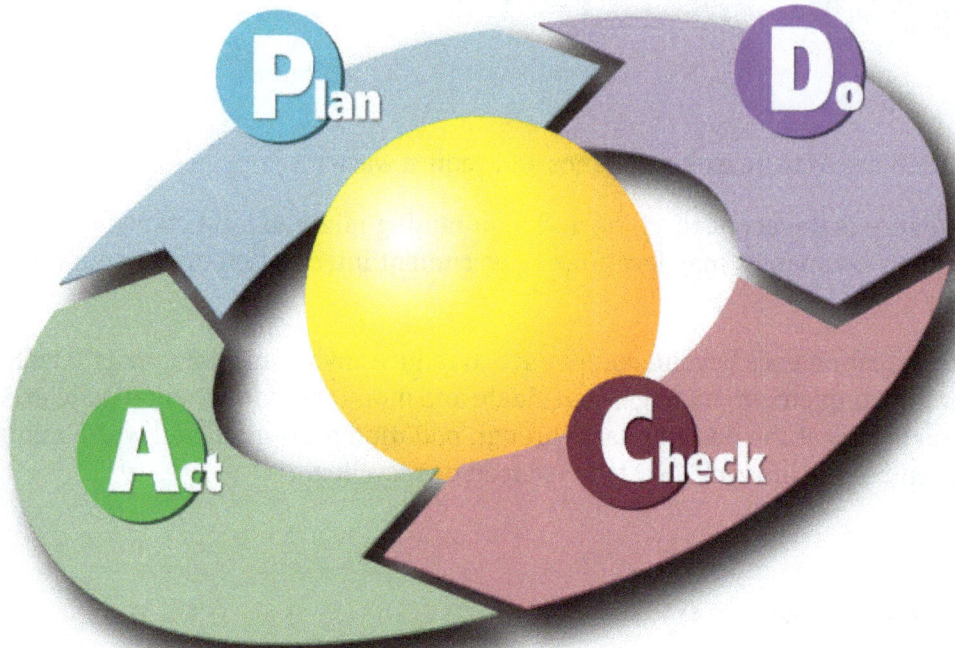

The PDCA cycle

An EMS follows a Plan-Do-Check-Act, or PDCA, Cycle. The diagram shows the process of first developing an environmental policy, planning the EMS, and then implementing it. The process also includes checking the system and acting on it. The model is continuous because an EMS is a process of continual improvement in which an organization is constantly reviewing and revising the system.

This is a model that can be used by a wide range of organizations — from manufacturing facilities to service industries to government agencies.

Other Meanings

An EMS can also be classified as

- a system which monitors, tracks and reports emissions information, particularly with respect to the oil and gas industry. EMSs are becoming web-based in response to the EPA's mandated greenhouse gas (GHG) reporting rule, which allows for reporting GHG emissions information via the internet.

- a centrally controlled and often automated network of devices (now frequently wireless using z-wave and zigbee technologies) used to control the internal environment of a building. Such a system namely acts as an interface between end user and energy (gas/electricity) consumption.

Life-Cycle Assessment

Life Cycle Assessment Overview

Life-cycle assessment (LCA, also known as life-cycle analysis, ecobalance, and cradle-to-grave analysis) is a technique to assess environmental impacts associated with all the stages of a product's life from cradle to grave (i.e., from raw material extraction through materials processing, manufacture, distribution, use, repair and maintenance, and disposal or recycling). LCAs can help avoid a narrow outlook on environmental concerns by:

- Compiling an inventory of relevant energy and material inputs and environmental releases;

- Evaluating the potential impacts associated with identified inputs and releases;

- Interpreting the results to help make a more informed decision.

Goals and Purpose

The goal of LCA is to compare the full range of environmental effects assignable to products and services by quantifying all inputs and outputs of material flows and assessing how these material flows affect the environment. This information is used to improve processes, support policy and provide a sound basis for informed decisions.

Life Cycle Assessment A systematic set of procedures for compiling and examining the inputs and outputs of materials and energy and the associated environmental impacts directly attributable to the functioning of a product or service system throughout its life cycle.

The term *life cycle* refers to the notion that a fair, holistic assessment requires the assessment of raw-material production, manufacture, distribution, use and disposal including all intervening transportation steps necessary or caused by the product's existence.

There are two main types of LCA. Attributional LCAs seek to establish (or attribute) the burdens associated with the production and use of a product, or with a specific service or process, at a point in time (typically the recent past). Consequential LCAs seek to identify the environmental consequences of a decision or a proposed change in a system under study (oriented to the future), which means that market and economic implications of a decision may have to be taken into account. Social LCA is under development as a different approach to life cycle thinking intended to assess social implications or potential impacts. Social LCA should be considered as an approach that is complementary to environmental LCA.

The procedures of life cycle assessment (LCA) are part of the ISO 14000 environmental management standards: in ISO 14040:2006 and 14044:2006. (ISO 14044 replaced earlier versions of ISO 14041 to ISO 14043.) GHG product life cycle assessments can also comply with specifications such as PAS 2050 and the GHG Protocol Life Cycle Accounting and Reporting Standard.

Four Main Phases

According to the ISO 14040 and 14044 standards, a Life Cycle Assessment is carried out in four distinct phases as illustrated in the figure shown to the right. The phases are often interdependent in that the results of one phase will inform how other phases are completed.

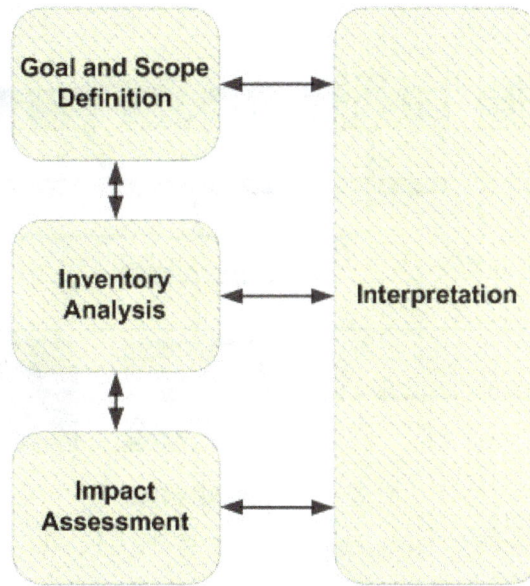

Illustration of LCA phases

Goal and Scope

An LCA starts with an explicit statement of the goal and scope of the study, which sets out the context of the study and explains how and to whom the results are to be communicated. This is a key step and the ISO standards require that the goal and scope of an LCA be clearly defined and consistent with the intended application. The goal and scope document therefore includes technical details that guide subsequent work:

- the functional unit, which defines what precisely is being studied and quantifies the service delivered by the product system, providing a reference to which the inputs and outputs can be related. Further, the functional unit is an important basis that enables alternative goods, or services, to be compared and analyzed. So to explain this a functional system which is inputs, processes and outputs contains a functional unit, that fulfills a function, for example paint is covering a wall, making a functional unit of 1m² covered for 10 years. The functional flow would be the items necessary for that function, so this would be a brush, tin of paint and the paint itself.

- the system boundaries;

- any assumptions and limitations;

- the allocation methods used to partition the environmental load of a process when several products or functions share the same process; allocation is commonly dealt with in one of three ways: system expansion, substitution and partition. Doing this is not easy and different methods may give different results

and

- the impact categories chosen for example human toxicity, smog, global warming, eutrophication.

Life Cycle Inventory

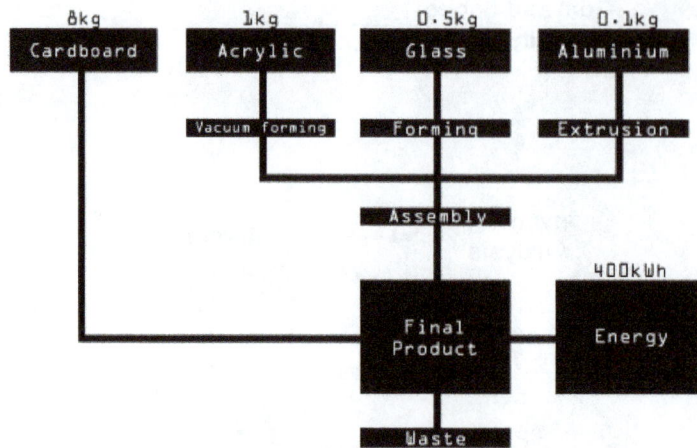

This is an example of a Life-cycle inventory (LCI) diagram

Life Cycle Inventory (LCI) analysis involves creating an inventory of flows from and to nature for a product system. Inventory flows include inputs of water, energy, and raw materials, and releases to air, land, and water. To develop the inventory, a flow model of the technical system is constructed using data on inputs and outputs. The flow model is typically illustrated with a flow chart that includes the activities that are going to be assessed in the relevant supply chain and gives a clear picture of the technical system boundaries. The input and output data needed for the construction of the model are collected for all activities within the system boundary, including from the supply chain (referred to as inputs from the techno-sphere).

The data must be related to the functional unit defined in the goal and scope definition. Data can be presented in tables and some interpretations can be made already at this stage. The results of the inventory is an LCI which provides information about all inputs and outputs in the form of elementary flow to and from the environment from all the unit processes involved in the study.

Inventory flows can number in the hundreds depending on the system boundary. For product LCAs at either the generic (i.e., representative industry averages) or brand-specific level, that data is typically collected through survey questionnaires. At an industry level, care has to be taken to ensure that questionnaires are completed by a representative sample of producers, leaning toward neither the best nor the worst, and fully representing any regional differences due to energy use, material sourcing or other factors. The questionnaires cover the full range of inputs and outputs, typically aiming to account for 99% of the mass of a product, 99% of the energy used in its production and any environmentally sensitive flows, even if they fall within the 1% level of inputs.

One area where data access is likely to be difficult is flows from the techno-sphere. The techno-sphere is more simply defined as the man-made world. Considered by geologists as secondary resources, these resources are in theory 100% recyclable; however, in a practical sense the primary goal is salvage. For an LCI, these technosphere products (supply chain products) are those that have been produced by man and unfortunately those completing a questionnaire about a process which uses man-made product as a means to an end will be unable to specify how much of a given input they use. Typically, they will not have access to data concerning inputs and outputs for previous production processes of the product. The entity undertaking the LCA must then turn to secondary sources if it does not already have that data from its own previous studies. National databases or data sets that come with LCA-practitioner tools, or that can be readily accessed, are the usual sources for that information. Care must then be taken to ensure that the secondary data source properly reflects regional or national conditions.

Life Cycle Impact Assessment

Inventory analysis is followed by impact assessment. This phase of LCA is aimed at evaluating the significance of potential environmental impacts based on the LCI flow results. Classical life cycle impact assessment (LCIA) consists of the following mandatory elements:

- selection of impact categories, category indicators, and characterization models;

- the classification stage, where the inventory parameters are sorted and assigned to specific impact categories; and

- impact measurement, where the categorized LCI flows are characterized, using one of many possible LCIA methodologies, into common equivalence units that are then summed to provide an overall impact category total.

In many LCAs, characterization concludes the LCIA analysis; this is also the last compulsory stage according to ISO 14044:2006. However, in addition to the above mandatory LCIA steps, other optional LCIA elements – normalization, grouping, and weighting – may be conducted depending on the goal and scope of the LCA study. In normalization, the results of the impact categories from the study are usually compared with the total impacts in the region of interest, the U.S. for example. Grouping consists of sorting and possibly ranking the impact categories. During weighting, the different environmental impacts are weighted relative to each other so that they can then be summed to get a single number for the total environmental impact. ISO 14044:2006 generally advises against weighting, stating that "weighting, shall not be used in LCA studies intended to be used in comparative assertions intended to be disclosed to the public". This advice is often ignored, resulting in comparisons that can reflect a high degree of subjectivity as a result of weighting.

Interpretation

Life Cycle Interpretation is a systematic technique to identify, quantify, check, and evaluate information from the results of the life cycle inventory and/or the life cycle impact assessment. The results from the inventory analysis and impact assessment are summarized during the interpretation phase. The outcome of the interpretation phase is a set of conclusions and recommendations for the study. According to ISO 14040:2006, the interpretation should include:

- identification of significant issues based on the results of the LCI and LCIA phases of an LCA;

- evaluation of the study considering completeness, sensitivity and consistency checks; and

- conclusions, limitations and recommendations.

A key purpose of performing life cycle interpretation is to determine the level of confidence in the final results and communicate them in a fair, complete, and accurate manner. Interpreting the results of an LCA is not as simple as "3 is better than 2, therefore Alternative A is the best choice"! Interpreting the results of an LCA starts with understanding the accuracy of the results, and ensuring they meet the goal of the study. This is accomplished by identifying the data elements that contribute significantly to each impact category, evaluating the sensitivity of these significant data elements, assessing the completeness and consistency of the study, and drawing conclusions and recommendations based on a clear understanding of how the LCA was conducted and the results were developed.

Reference Test

More specifically, the best alternative is the one that the LCA shows to have the least cradle-to-grave environmental negative impact on land, sea, and air resources.

LCA Uses

Based on a survey of LCA practitioners carried out in 2006 LCA is mostly used to support business strategy (18%) and R&D (18%), as input to product or process design (15%), in education (13%) and for labeling or product declarations (11%). LCA will be continuously integrated into the built environment as tools such as the European ENSLIC Building project guidelines for buildings or developed and implemented, which provide practitioners guidance on methods to implement LCI data into the planning and design process.

Major corporations all over the world are either undertaking LCA in house or commissioning studies, while governments support the development of national databases to support LCA. Of particular note is the growing use of LCA for ISO Type III labels called Environmental Product Declarations, defined as "quantified environmental data for a product with pre-set categories of parameters based on the ISO 14040 series of standards, but not excluding additional environmental information". These third-party certified LCA-based labels provide an increasingly important basis for assessing the relative environmental merits of competing products. Third-party certification plays a major role in today's industry. Independent certification can show a company's dedication to safer and environmental friendlier products to customers and NGOs.

LCA also has major roles in environmental impact assessment, integrated waste management and pollution studies.

Data Analysis

A life cycle analysis is only as valid as its data; therefore, it is crucial that data used for the completion of a life cycle analysis are accurate and current. When comparing different life cycle analyses with one another, it is crucial that equivalent data are available for both products or processes in question. If one product has a much higher availability of data, it cannot be justly compared to another product which has less detailed data.

There are two basic types of LCA data – unit process data and environmental input-output data (EIO), where the latter is based on national economic input-output data. Unit process data are derived from direct surveys of companies or plants producing the product of interest, carried out at a unit process level defined by the system boundaries for the study.

Data validity is an ongoing concern for life cycle analyses. Due to globalization and the rapid pace of research and development, new materials and manufacturing methods are continually being introduced to the market. This makes it both very important and very difficult to use up-to-date information when performing an LCA. If an LCA's conclusions are to be valid, the data must be recent; however, the data-gathering process takes time. If a product and its related processes have not undergone significant revisions since the last LCA data was collected, data validity is not a problem. However, consumer electronics such as cell phones can be redesigned as often as every 9 to 12 months, creating a need for ongoing data collection.

The life cycle considered usually consists of a number of stages including: materials extraction, processing and manufacturing, product use, and product disposal. If the most environmentally harmful of these stages can be determined, then impact on the environment can be efficiently reduced by focusing on making changes for that particular phase. For example, the most energy-intensive life phase of an airplane or car is during use due to fuel consumption. One of the most effective ways to increase fuel efficiency is to decrease vehicle weight, and thus, car and airplane manufacturers can decrease environmental impact in a significant way by replacing heavier materials with lighter ones such as aluminium or carbon fiber-reinforced elements. The reduction during the use phase should be more than enough to balance additional raw material or manufacturing cost.

Variants

Cradle-to-Grave

Cradle-to-grave is the full Life Cycle Assessment from resource extraction ('cradle') to use phase and disposal phase ('grave'). For example, trees produce paper, which can be recycled into low-energy production cellulose (fiberised paper) insulation, then used as an energy-saving device in the ceiling of a home for 40 years, saving 2,000 times the fossil-fuel energy used in its production. After 40 years the cellulose fibers are replaced and the old fibers are disposed of, possibly incinerated. All inputs and outputs are considered for all the phases of the life cycle.

Cradle-to-Gate

Cradle-to-gate is an assessment of a *partial* product life cycle from resource extraction (*cradle*) to the factory gate (i.e., before it is transported to the consumer). The use phase and disposal phase of the product are omitted in this case. Cradle-to-gate assessments are sometimes the basis for environmental product declarations (EPD) termed business-to-business EDPs. One of the significant uses of the cradle-to-gate approach compiles the life cycle inventory (LCI) using cradle-to-gate. This allows the LCA to collect all of the impacts leading up to resources being purchased by the facility. They can then add the steps involved in their transport to plant and manufacture process to more easily produce their own cradle-to-gate values for their products.

Cradle-to-Cradle or Closed Loop Production

Cradle-to-cradle is a specific kind of cradle-to-grave assessment, where the end-of-life disposal step for the product is a recycling process. It is a method used to minimize the environmental impact of products by employing sustainable production, operation, and disposal practices and aims to incorporate social responsibility into product development. From the recycling process originate new, identical products (e.g., asphalt pavement from discarded asphalt pavement, glass bottles from collected glass bottles), or different products (e.g., glass wool insulation from collected glass bottles).

Allocation of burden for products in open loop production systems presents considerable challenges for LCA. Various methods, such as the avoided burden approach have been proposed to deal with the issues involved.

Gate-to-Gate

Gate-to-gate is a partial LCA looking at only one value-added process in the entire production chain. Gate-to-gate modules may also later be linked in their appropriate production chain to form a complete cradle-to-gate evaluation.

Well-to-Wheel

Well-to-wheel is the specific LCA used for transport fuels and vehicles. The analysis is often broken down into stages entitled "well-to-station", or "well-to-tank", and "station-to-wheel" or "tank-to-wheel", or "plug-to-wheel". The first stage, which incorporates the feedstock or fuel production and processing and fuel delivery or energy transmission, and is called the "upstream" stage, while the stage that deals with vehicle operation itself is sometimes called the "downstream" stage. The well-to-wheel analysis is commonly used to assess total energy consumption, or the energy conversion efficiency and emissions impact of marine vessels, aircraft and motor vehicles, including their carbon footprint, and the fuels used in each of these transport modes.

The well-to-wheel variant has a significant input on a model developed by the Argonne National Laboratory. The Greenhouse gases, Regulated Emissions, and Energy use in Transportation (GREET) model was developed to evaluate the impacts of new fuels and vehicle technologies. The model evaluates the impacts of fuel use using a well-to-wheel evaluation while a traditional cradle-to-grave approach is used to determine the impacts from the vehicle itself. The model reports energy use, greenhouse gas emissions, and six additional pollutants: volatile organic compounds (VOCs), carbon monoxide (CO), nitrogen oxide (NOx), particulate matter with size smaller than 10 micrometre (PM10), particulate matter with size smaller than 2.5 micrometre (PM2.5), and sulfur oxides (SOx).

Economic Input–Output Life Cycle Assessment

Economic input–output LCA (EIOLCA) involves use of aggregate sector-level data on how much environmental impact can be attributed to each sector of the economy and how much each sector purchases from other sectors. Such analysis can account for long chains (for example, building an automobile requires energy, but producing energy requires vehicles, and building those vehicles requires energy, etc.), which somewhat alleviates the scoping problem of process LCA; however, EIOLCA relies on sector-level averages that may or may not be representative of the specific subset of the sector relevant to a particular product and therefore is not suitable for evaluating the environmental impacts of products. Additionally the translation of economic quantities into environmental impacts is not validated.

Ecologically Based LCA

While a conventional LCA uses many of the same approaches and strategies as an Eco-LCA, the latter considers a much broader range of ecological impacts. It was designed to provide a guide to wise management of human activities by understanding the direct and indirect impacts on ecological resources and surrounding ecosystems. Developed by Ohio State University Center for resilience, Eco-LCA is a methodology that quantitatively takes into account regulating and supporting services during the life cycle of economic goods and products. In this approach services are categorized in four main groups: supporting, regulating, provisioning and cultural services.

Life Cycle Energy Analysis

Life cycle energy analysis (LCEA) is an approach in which all energy inputs to a product are accounted for, not only direct energy inputs during manufacture, but also all energy inputs needed to produce components, materials and services needed for the manufacturing process. An earlier term for the approach was *energy analysis*.

With LCEA, the *total life cycle energy input* is established.

Energy Production

It is recognized that much energy is lost in the production of energy commodities themselves, such as nuclear energy, photovoltaic electricity or high-quality petroleum products. *Net energy content* is the energy content of the product minus energy input used during extraction and conversion, directly or indirectly. A controversial early result of LCEA claimed that manufacturing solar cells requires more energy than can be recovered in using the solar cell. The result was refuted. Another new concept that flows from life cycle assessments is Energy Cannibalism. Energy Cannibalism refers to an effect where rapid growth of an entire energy-intensive industry creates a need for energy that uses (or cannibalizes) the energy of existing power plants. Thus during rapid growth the industry as a whole produces no energy because new energy is used to fuel the embodied energy of future power plants. Work has been undertaken in the UK to determine the life cycle energy (alongside full LCA) impacts of a number of renewable technologies.

Energy Recovery

If materials are incinerated during the disposal process, the energy released during burning can be harnessed and used for electricity production. This provides a low-impact energy source, especially when compared with coal and natural gas While incineration produces more greenhouse gas emissions than landfilling, the waste plants are well-fitted with filters to minimize this negative impact. A recent study comparing energy consumption and greenhouse gas emissions from landfilling (without energy recovery) against incineration (with energy recovery) found incineration to be superior in all cases except for when landfill gas is recovered for electricity production.

Criticism

A criticism of LCEA is that it attempts to eliminate monetary cost analysis, that is replace the currency by which economic decisions are made with an energy currency. It has also been argued that energy efficiency is only one consideration in deciding which alternative process to employ, and that it should not be elevated to the only criterion for determining environmental acceptability; for example, simple energy analysis does not take into account the renewability of energy flows or the toxicity of waste products; however the life cycle assessment does help companies become more familiar with environmental properties and improve their environmental system. Incorporating Dynamic LCAs of renewable energy technologies (using sensitivity analyses to project future improvements in renewable systems and their share of the power grid) may help mitigate this criticism.

In recent years, the literature on life cycle assessment of energy technology has begun to reflect the interactions between the current electrical grid and future energy technology. Some papers have focused on energy life cycle, while others have focused on carbon dioxide (CO_2) and other greenhouse gases. The essential critique given by these sources is that when considering energy technology, the growing nature of the power grid must be taken into consideration. If this is not done, a given class of energy technology may emit more CO_2 over its lifetime than it mitigates.

A problem the energy analysis method cannot resolve is that different energy forms (heat, electricity, chemical energy etc.) have different quality and value even in natural sciences, as a consequence of the two main laws of thermodynamics. A thermodynamic measure of the quality of energy is exergy. According to the first law of thermodynamics, all energy inputs should be accounted with equal weight, whereas by the second law diverse energy forms should be accounted by different values.

The conflict is resolved in one of these ways:

- value difference between energy inputs is ignored,

- a value ratio is arbitrarily assigned (e.g., a joule of electricity is 2.6 times more valuable than a joule of heat or fuel input),

- the analysis is supplemented by economic (monetary) cost analysis,

- exergy instead of energy can be the metric used for the life cycle analysis.

Critiques

Life cycle assessment is a powerful tool for analyzing commensurable aspects of quantifiable systems. Not every factor, however, can be reduced to a number and inserted into a model. Rigid system boundaries make accounting for changes in the system difficult. This is sometimes referred to as the boundary critique to systems thinking. The accuracy and availability of data can also contribute to inaccuracy. For instance, data from generic processes may be based on averages, unrepresentative sampling, or outdated results. Additionally, social implications of products are generally lacking in LCAs. Comparative life-cycle analysis is often used to determine a better process or product to use. However, because of aspects like differing system boundaries, different statistical information, different product uses, etc., these studies can easily be swayed in favor of one product or process over another in one study and the opposite in another study based on varying parameters and different available data. There are guidelines to help reduce such conflicts in results but the method still provides a lot of room for the researcher to decide what is important, how the product is typically manufactured, and how it is typically used.

An in-depth review of 13 LCA studies of wood and paper products found a lack of consistency in the methods and assumptions used to track carbon during the product lifecycle. A wide variety of methods and assumptions were used, leading to different and potentially contrary conclusions – particularly with regard to carbon sequestration and methane generation in landfills and with carbon accounting during forest growth and product use.

Streamline LCA

This process includes three steps. First, a proper method should be selected to combine adequate accuracy with acceptable cost burden in order to guide decision making. Actually, in LCA process, besides streamline LCA, Eco-screening and complete LCA are usually considered as well. However, the former one only could provide limited details and the latter one with more detailed information is more expensive. Second, single measure of stress should be selected. Typical LCA output includes resource consumption, energy consumption, water consumption, emission of CO_2, toxic residues and so on. One of these outputs is used as the main factor to measure in streamline LCA. Energy consumption and CO_2 emission are often regarded as "practical indicators". Last, stress selected in step 2 is used as standard to assess phase of life separately and identify the most damaging phase. For instance, for a family car, energy consumption could be used as the single stress factor to assess each phase of life. The result shows that the most energy intensive phase for a family car is the usage stage.

• Agroecology	
• Agroecosystem analysis	• Ecodesign
• Anthropogenic metabolism	• End-of-life (product)
• Biofuel	• Environmental pricing reform
• Carbon footprint	• Greenhouse gas
• Circular Economy	• GREET Model
• Cradle to Cradle	• Industrial ecology
• Depreciation	• ISO 15686
• Design for Environment	• Industrial metabolism
• Dimension stone Stone: life-cycle assessment and best practices	• Whole-life cost

References

- Samuel Bowles: Microeconomics: Behavior, Institutions, and Evolution, Princeton University Press, pp. 27–29 (2004) ISBN 0-691-09163-3

- Leakey, Richard and Roger Lewin, 1996, *The Sixth Extinction : Patterns of Life and the Future of Humankind*, Anchor, ISBN 0-385-46809-1

- *Benkler, Yochai (2006). The Wealth of Networks: How Social Production Transforms Markets and Freedom. New Haven, Conn: Yale University Press. p. 3. ISBN 0-300-11056-1.*

- *S. Singh; B. R. Bakshi (2009). "Eco-LCA: A Tool for Quantifying the Role of Ecological Resources in LCA". International Symposium on Sustainable Systems and Technology: 1–6. doi:10.1109/ISSST.2009.5156770. ISBN 978-1-4244-4324-6.*

- Hendrickson, C. T., Lave, L. B., and Matthews, H. S. (2005). *Environmental Life Cycle Assessment of Goods and Services: An Input–Output Approach*, Resources for the Future Press ISBN 1-933115-24-6.

- *Gorriz-Misfud, Elena; Secco, L; Pisani, E (2016). "Exploring the interlinkages between governance and social capital: A dynamic model for forestry.". Forest Policy and Economics. doi:10.1016/j.forpol.2016.01.006.*

- *Lloyd, William Forster (1833). Two lectures on the checks to population. England: Oxford University. Retrieved 2016-03-13.*

- Ibrahim, Ahmed (2015): The tragedy of the commons and prisoner's dilemma may improve our realization of the theory of life and provide us with advanced therapeutic ways. figshare.

- *de Groot, Rudolf; Matthew Wilson; Roelof Boumans (2002). "A typology for the classification, description and valuation of ecosystem functions, goods and services" (PDF). Ecological Economics. **41**: 393–408. doi:10.1016/s0921-8009(02)00089-7. Retrieved 2015-03-02.*

- *Prause, Christian (September 5, 2011). "Reputation-based self-management of software process artifact quality in consortium research projects". ACM. Retrieved 22 October 2013.*

- *"A Smiley Face Emoticon For Your Electric Bill | Unambiguously Ambidextrous". Unambig.com. Archived from the original on 2011-08-31. Retrieved 22 October 2013.*

- "PAS 2050:2011 Specification for the assessment of the life cycle greenhouse gas emissions of goods and services". *BSI*. Retrieved on: 25 April 2013.

- *Brinkman, Norman; Eberle, Ulrich; Formanski, Volker; Grebe, Uwe-Dieter; Matthe, Roland (15 April 2012). "Vehicle Electrification - Quo Vadis". VDI. Retrieved 27 April 2013.*

- Jaeger, William. *Environmental Economics for Tree Huggers and Other Skeptics*, p. 80 (Island Press 2012): "Economists often say that externalities need to be 'internalized,' meaning that some action needs to be taken to correct this kind of market failure."

- *Steven Johnson (September 21, 2012). "The Internet? We Built That". New York Times. Retrieved 2012-09-24. The Harvard legal scholar Yochai Benkler has called this phenomenon 'commons-based peer production'.*

- Daniel, T. C. et al. 2012: *Contributions of cultural services to the ecosystem services agenda*. Proc. Natl. Acad. Sci. USA 109: 8812–8819 [2].

4

Environmental Laws and Environmental Policies

This chapter specifically deals with the laws and regulations used worldwide to conserve environment and preserve ecosystems. The particular topics discussed in this chapter are laws, justice, policy and governance with respect to environmental protection. It also glances upon globalization and environmental movement. This chapter will elaborate the various issues and problems faced by authorities while implementing these laws.

Environmental Policy

Environmental policy refers to the commitment of an organization to the laws, regulations, and other policy mechanisms concerning environmental issues. These issues generally include air and water pollution, solid waste management, ecosystem management, maintenance of biodiversity, the protection of natural resources, wildlife and endangered species. Policies concerning energy or regulation of toxic substances including pesticides and many types of industrial waste are part of the topic of environmental policy. This policy can be deliberately taken to direct and oversee human activities and thereby prevent harmful effects on the biophysical environment and natural resources, as well as to make sure that changes in the environment do not have harmful effects on humans.

Definition

It is useful to consider that environmental policy comprises two major terms: environment and policy. Environment refers to the physical ecosystems, but can also take into consideration the social dimension (quality of life, health) and an economic dimension (resource management, biodiversity). Policy can be defined as a "course of action or principle adopted or proposed by a government, party, business or individual". Thus, environmental policy focuses on problems arising from human impact on the environment, which retroacts onto human society by having a (negative) impact on human values such as good health or the 'clean and green' environment.

Environmental issues generally addressed by environmental policy include (but are not limited to) air and water pollution, waste management, ecosystem management, biodiversity protection, the protection of natural resources, wildlife and endangered species, and the preservation of these natural resources for future generations. Relatively recently, environmental policy has also attended to the communication of environmental issues.

Rationale

The rationale for governmental involvement in the environment is market failure in the form of forces beyond the control of one person, including the free rider problem and the tragedy of the commons. An example of an externality is when a factory produces waste pollution which may be dumped into a river, ultimately contaminating water. The cost of such action is paid by society-at-large, when they must clean the water before drinking it and is external to the costs of the factory. The free rider problem is when the private marginal cost of taking action to protect the environment is greater than the private marginal benefit, but the social marginal cost is less than the social marginal benefit. The tragedy of the commons is the problem that, because no one person owns the commons, each individual has an incentive to utilize common resources as much as possible. Without governmental involvement, the commons is overused. Examples of tragedies of the commons are overfishing and overgrazing.

Instruments, Problems and Issues

Environmental policy instruments are tools used by governments to implement their environmental policies. Governments may use a number of different types of instruments. For example, economic incentives and market-based instruments such as taxes and tax exemptions, tradable permits, and fees can be very effective to encourage compliance with environmental policy.

Bilateral agreements between the government and private firms and commitments made by firms independent of government requirement are examples of voluntary environmental measures. Another instrument is the implementation of greener public purchasing programs.

Several instruments are sometimes combined in a policy mix to address a certain environmental problem. Since environmental issues have many aspects, several policy instruments may be needed to adequately address each one. Furthermore, a combination of different policies may give firms greater flexibility in policy compliance and reduce uncertainty as to the cost of such compliance.

Government policies must be carefully formulated so that the individual measures do not undermine one another, or create a rigid and cost-ineffective framework. Overlapping policies result in unnecessary administrative costs, increasing the cost of implementation. To help governments realize their policy goals, the OECD Environment Directorate collects data on the efficiency and consequences of environmental policies implemented by the national governments. The website, www.economicinstruments.com, provides database detailing countries' experiences with their environmental policies. The United Nations Economic Commission for Europe, through UNECE Environmental Performance Reviews, evaluates progress made by its member countries in improving their environmental policies.

The current reliance on a market-based framework is controversial, however, and many environmentalists contend that a more radical, overarching approach is needed than a set of specific initiatives, to deal with the climate change. For example, energy efficiency measures may actually increase energy consumption in the absence of a cap on fossil fuel use, as people might drive more fuel-efficient cars. Thus, Aubrey Meyer calls for a 'framework-based market' of contraction and convergence. The Cap and Share and the Sky Trust are proposals based on the idea.

Environmental impact assessments (EIA) are conducted to compare impacts of various policy alternatives. Moreover, it is assumed that policymakers make rational decisions based on the merits of the project. Eccleston and March argue that although policymakers normally have access to reasonably accurate information, political and economic factors often lead to environmentally destructive decisions in the long run.

The decision-making theory casts doubt on this premise. Irrational decisions are reached based on unconscious biases, illogical assumptions, and the desire to avoid ambiguity and uncertainty.

Eccleston identifies and describes five of the most critical environmental policy issues facing humanity: water scarcity, food scarcity, climate change, the peak oil, and the population paradox.

Research and Innovation Policy

Synergic to the environmental policy is the environmental research and innovation policy. An example is the European environmental research and innovation policy, which aims at defining and implementing a transformative agenda to greening the economy and the society as a whole so to achieve a truly sustainable development. Europe is particularly active in this field, via a set of strategies, actions and programmes to promote more and better research and innovation for building a resource-efficient, climate resilient society and thriving economy in sync with its natural environment. Research and innovation in Europe are financially supported by the programme Horizon 2020, which is also open to participation worldwide.

History

The 1960s marked the beginning of modern environmental policy making. Although mainstream America remained oblivious to environmental concerns, the stage had been set for change by the publication of Rachel Carson's New York Times bestseller Silent Spring in 1962. Earth Day founder Gaylord Nelson, then a U.S. Senator from Wisconsin, after witnessing the ravages of the 1969 massive oil spill in Santa Barbara, California. Administrator Ruckelshaus was confirmed by the Senate on December 2, 1970, which is the traditional date used as the birth of the agency. Five months earlier, in July 1970, President Nixon had signed Reorganization Plan No. 3 calling for the establishment of EPA in July 1970. At the time, Environmental Policy was a bipartisan issue and the efforts of the United States of America helped spark countries around the world to create environmental policies. During this period, legislation was passed to regulate pollutants that go into the air, water tables, and solid waste disposal. President Nixon signed the Clean Air Act in 1970 which set the USA as one of the world leaders in environmental conservation.

In the European Union, the very first Environmental Action Programmed was adopted by national government representatives in July 1973 during the first meeting of the Council of Environmental Ministers. Since then an increasingly dense network of legislation has developed, which now extends to all areas of environmental protection including air pollution control, water protection and waste policy but also nature conservation and the control of chemicals, biotechnology and other industrial risks. EU environmental policy has thus become a core area of European politics.

Overall organizations are becoming more aware of their environmental risks and performance requirements. In line with the ISO 14001 standard they are developing environmental policies suitable for their organization. This statement outlines environmental performance of the organization as well as its environmental objectives. Written by top management of the organization they document a commitment to continuous improvement and complying with legal and other requirements, such as the environmental policy objectives set by their governments.

Environmental Policy Integration

The concept of environmental policy integration (EPI) refers to the process of integrating environmental objectives into non-environmental policy areas, such as energy, agriculture and transport, rather than leaving them to be pursued solely through purely environmental policy practices. This is oftentimes particularly challenging because of the need to reconcile global objectives and international rules with domestic needs and laws. EPI is widely recognised as one of the key elements of sustainable development. More recently, the notion of 'climate policy integration', also denoted as 'mainstreaming', has been applied to indicate the integration of climate considerations (both mitigation and adaptation) into the normal (often economically focused) activity of government.

Environmental Policy Studies

Given the growing need for trained environmental practitioners, graduate schools throughout the world offer specialized professional degrees in environmental policy studies. While there is not a standard curriculum, students typically take classes in policy analysis, environmental science, environmental law and politics, ecology, energy, and natural resource management. Graduates of these programs are employed by governments, international organizations, private sector, think tanks, universities, and so on.

Due to the lack of standard nomenclature, institutions use varying designations to refer to academic degrees they award. However, the degrees typically fall in one of four broad categories: master of arts, master of science, master of public administration, and PhD in environmental policy. Sometimes, more specific names are used to reflect the focus of the academic program. For example, the Middlebury Institute of International Studies at Monterey awards master of arts in international environmental policy (MAIEP) to emphasize the international orientation of the curriculum.

Environmental Law

Environmental law - or "environmental and natural resources law" - is a collective term describing the network of treaties, statutes, regulations, and common and customary laws addressing the effects of human activity on the natural environment.

Regulatory Subjects

The broad category of "environmental law" may be broken down into a number of more specific regulatory subjects. While there is no single agreed-upon taxonomy, the core environmental law regimes address environmental pollution. A related but distinct set of regulatory regimes, now strongly influenced by environmental legal principles, focus on the management of specific natural resources, such as forests, minerals, or fisheries. Other areas, such as environmental impact assessment, may not fit neatly into either category, but are nonetheless important components of environmental law.

Impact Assessment

Environmental impact assessment (EA) is the term used for the assessment of the environmental consequences (positive and negative) of a plan, policy, program, or project prior to the decision to move forward with the proposed action. In this context, the term 'environmental impact assessment' (EIA) is usually used when applied to concrete projects and the term 'strategic environmental assessment' applies to policies, plans and programmes (Fischer, 2016). Environmental assessments may be governed by rules of administrative procedure regarding public participation and documentation of decision making, and may be subject to judicial review.

Air Quality

Industrial air pollution now regulated by air quality law.

Air quality laws govern the emission of air pollutants into the atmosphere. A specialized subset of air quality laws regulate the quality of air inside buildings. Air quality laws are often designed specifically to protect human health by limiting or eliminating airborne pollutant concentrations. Other initiatives are designed to address broader ecological problems, such as limitations on chemicals that affect the ozone layer, and emissions trading programs to address acid rain or climate change. Regulatory efforts include identifying and categorizing air pollutants, setting limits on acceptable emissions levels, and dictating necessary or appropriate mitigation technologies.

Water Quality

A typical stormwater outfall, subject to water quality law.

Water quality laws govern the release of pollutants into water resources, including surface water, ground water, and stored drinking water. Some water quality laws, such as drinking water regulations, may be designed solely with reference to human health. Many others, including restrictions on the alteration of the chemical, physical, radiological, and biological characteristics of water resources, may also reflect efforts to protect aquatic ecosystems more broadly. Regulatory efforts may include identifying and categorizing water pollutants, dictating acceptable pollutant concentrations in water resources, and limiting pollutant discharges from effluent sources. Regulatory areas include sewage treatment and disposal, industrial and agricultural waste water management, and control of surface runoff from construction sites and urban environments.

Waste Management

A municipal landfill, operated pursuant to waste management law.

Waste management laws govern the transport, treatment, storage, and disposal of all manner of waste, including municipal solid waste, hazardous waste, and nuclear waste, among many other types. Waste laws are generally designed to minimize or eliminate the uncontrolled dispersal of waste materials into the environment in a manner that may cause ecological or biological harm, and include laws designed to reduce the generation of waste and promote or mandate waste recycling. Regulatory efforts include identifying and categorizing waste types and mandating transport, treatment, storage, and disposal practices.

Contaminant Cleanup

Oil spill emergency response, governed by environmental cleanup law.

Environmental cleanup laws govern the removal of pollution or contaminants from environmental media such as soil, sediment, surface water, or ground water. Unlike pollution control laws, cleanup laws are designed to respond after-the-fact to environmental contamination, and consequently must often define not only the necessary response actions, but also the parties who may be responsible for undertaking (or paying for) such actions. Regulatory requirements may include rules for emergency response, liability allocation, site assessment, remedial investigation, feasibility studies, remedial action, post-remedial monitoring, and site reuse.

Chemical Safety

Chemical safety laws govern the use of chemicals in human activities, particularly man-made chemicals in modern industrial applications. As contrasted with media-oriented environmental laws (e.g., air or water quality laws), chemical control laws seek to manage the (potential) pollutants themselves. Regulatory efforts include banning specific chemical constituents in consumer products (e.g., Bisphenol A in plastic bottles), and regulating pesticides.

Water Resources

An irrigation ditch, operated in accordance with water resources law.

Water resources laws govern the ownership and use of water resources, including surface water and ground water. Regulatory areas may include water conservation, use restrictions, and ownership regimes.

Mineral Resources

Mineral resource laws cover several basic topics, including the ownership of the mineral resource and who can work them. Mining is also affected by various regulations regarding the health and safety of miners, as well as the environmental impact of mining.

Forest Resources

A timber operation, regulated by forestry law.

Forestry laws govern activities in designated forest lands, most commonly with respect to forest management and timber harvesting. Ancillary laws may regulate forest land acquisition and prescribed burn practices. Forest management laws generally adopt management policies, such as multiple use and sustained yield, by which public forest resources are to be managed. Governmental agencies are generally responsible for planning and implementing forestry laws on public forest lands, and may be involved in forest inventory, planning, and conservation, and oversight of timber sales. Broader initiatives may seek to slow or reverse deforestation.

Wildlife and Plants

Wildlife laws govern the potential impact of human activity on wild animals, whether directly on individuals or populations, or indirectly via habitat degradation. Similar laws may operate to protect plant species. Such laws may be enacted entirely to protect biodiversity, or as a means for protecting species deemed important for other reasons. Regulatory efforts may including the creation of special conservation statuses, prohibitions on killing, harming, or disturbing protected species, efforts to induce and support species recovery, establishment of wildlife refuges to support conservation, and prohibitions on trafficking in species or animal parts to combat poaching.

Fish and Game

Fish and game laws regulate the right to pursue and take or kill certain kinds of fish and wild animal (game). Such laws may restrict the days to harvest fish or game, the number of animals caught per person, the species harvested, or the weapons or fishing gear used. Such laws may seek to balance dueling needs for preservation and harvest and to manage both environment and populations of fish and game. Game laws can provide a legal structure to collect license fees and other money which is used to fund conservation efforts as well as to obtain harvest information used in wildlife management practice.

Important Principles

Environmental law has developed in response to emerging awareness of and concern over issues impacting the entire world. While laws have developed piecemeal and for a variety of reasons, some effort has gone into identifying key concepts and guiding principles common to environmental law as a whole. The principles discussed below are not an exhaustive list and are not universally recognized or accepted. Nonetheless, they represent important principles for the understanding of environmental law around the world.

Sustainable Development

Defined by the United Nations Environment Programme as "development that meets the needs of the present without compromising the ability of future generations to meet their own needs," sustainable development may be considered together with the concepts of "integration" (development cannot be considered in isolation from sustainability) and "interdependence" (social and economic development, and environmental protection, are interdependent). Laws mandating environmental impact assessment and requiring or encouraging development to minimize environmental impacts may be assessed against this principle.

The modern concept of sustainable development was a topic of discussion at the 1972 United Nations Conference on the Human Environment (Stockholm Conference), and the driving force behind the 1983 World Commission on Environment and Development (WCED, or Bruntland Commission). In 1992, the first UN Earth Summit resulted in the Rio Declaration, Principle 3 of which reads: "The right to development must be fulfilled so as to equitably meet developmental and environmental needs of present and future generations." Sustainable development has been a core concept of international environmental discussion ever since, including at the World Summit on Sustainable Development (Earth Summit 2002), and the United Nations Conference on Sustainable Development (Earth Summit 2012, or Rio+20).

Equity

Defined by UNEP to include intergenerational equity - "the right of future generations to enjoy a fair level of the common patrimony" - and intragenerational equity - "the right of all people within the current generation to fair access to the current generation's entitlement to the Earth's natural resources" - environmental equity considers the present generation under an obligation to account for long-term impacts of activities, and to act to sustain the global environment and resource base for future generations. Pollution control and resource management laws may be assessed against this principle.

Transboundary Responsibility

Defined in the international law context as an obligation to protect one's own environment, and to prevent damage to neighboring environments, UNEP considers transboundary responsibility at the international level as a potential limitation on the rights of the sovereign state. Laws that act to limit externalities imposed upon human health and the environment may be assessed against this principle.

Public Participation and Transparency

Identified as essential conditions for "accountable governments . . ., industrial concerns," and organizations generally, public participation and transparency are presented by UNEP as requiring "effective protection of the human right to hold and express opinions and to seek, receive and impart ideas," "a right of access to appropriate, comprehensible and timely information held by governments and industrial concerns on economic and social policies regarding the sustainable use of natural resources and the protection of the environment, without imposing undue financial burdens upon the applicants and with adequate protection of privacy and business confidentiality," and "effective judicial and administrative proceedings." These principles are present in environmental impact assessment, laws requiring publication and access to relevant environmental data, and administrative procedure.

Precautionary Principle

One of the most commonly encountered and controversial principles of environmental law, the Rio Declaration formulated the precautionary principle as follows:

In order to protect the environment, the precautionary approach shall be widely applied by States according to their capabilities. Where there are threats of serious or irreversible damage, lack of full scientific certainty shall not be used as a reason for postponing cost-effective measures to prevent environmental degradation.

The principle may play a role in any debate over the need for environmental regulation.

Prevention

The concept of prevention . . . can perhaps better be considered an overarching aim that gives rise to a multitude of legal mechanisms, including prior assessment of environmental harm, licensing or authorization that set out the conditions for operation and the consequences for violation of the conditions, as well as the adoption of strategies and policies. Emission limits and other product or process standards, the use of best available techniques and similar techniques can all be seen as applications of the concept of prevention.

Polluter Pays Principle

The polluter pays principle stands for the idea that "the environmental costs of economic activities, including the cost of preventing potential harm, should be internalized rather than imposed upon society at large." All issues related to responsibility for cost for environmental remediation and compliance with pollution control regulations involve this principle.

History

Early examples of legal enactments designed to consciously preserve the environment, for its own sake or human enjoyment, are found throughout history. In the common law, the primary protection was found in the law of nuisance, but this only allowed for private actions for damages or injunctions if there was harm to land. Thus smells emanating from pig stys, strict liability against dumping rubbish, or damage from exploding dams. Private enforcement, however, was limited and found to be woefully inadequate to deal with major environmental threats, particularly threats to common resources. During the "Great Stink" of 1858, the dumping of sewerage into the River Thames began to smell so ghastly in the summer heat that Parliament had to be evacuated. Ironically, the Metropolitan Commission of Sewers Act 1848 had allowed the Metropolitan Commission for Sewers to close cesspits around the city in an attempt to "clean up" but this simply led people to pollute the river. In 19 days, Parliament passed a further Act to build the London sewerage system. London also suffered from terrible air pollution, and this culminated in the "Great Smog" of 1952, which in turn triggered its on legislative response: the Clean Air Act 1956. The basic regulatory structure was to set limits on emissions for households and business (particularly burning coal) while an inspectorate would enforce compliance.

Notwithstanding early analogues, the concept of "environmental law" as a separate and distinct body of law is a twentieth-century development. The recognition that the natural environment was fragile and in need of special legal protections, the translation of that recognition into legal structures, the development of those structures into a larger body of "environmental law," and the strong influence of environmental law on natural resource laws, did not occur until about the 1960s. At that time, numerous influences - including a growing awareness of the unity and fragility of the biosphere; increased public concern over the impact of industrial activity on natural resources and human health; the increasing strength of the regulatory state; and more broadly the advent and success of environmentalism as a political movement - coalesced to produce a huge new body of law in a relatively short period of time. While the modern history of environmental law is one of continuing controversy, by the end of the twentieth century environmental law had been established as a component of the legal landscape in all developed nations of the world, many developing ones, and the larger project of international law.

Controversy

Environmental law is a continuing source of controversy. Debates over the necessity, fairness, and cost of environmental regulation are ongoing, as well as regarding the appropriateness of regulations vs. market solutions to achieve even agreed-upon ends.

Allegations of scientific uncertainty fuel the ongoing debate over greenhouse gas regulation, and are a major factor in debates over whether to ban particular pesticides. In cases where the science is well-settled, it is not unusual to find that corporations intentionally hide or distort the facts, or sow confusion.

It is very common for regulated industry to argue against environmental regulation on the basis of cost. Difficulties arise in performing cost-benefit analysis of environmental issues. It is difficult to quantify the value of an environmental value such as a healthy ecosystem, clean air, or species diversity. Many environmentalists' response to pitting economy vs. ecology is summed up by former Senator and founder of Earth Day Gaylord Nelson, "The economy is a wholly owned subsidiary of the environment, not the other way around." Furthermore, environmental issues are seen by many as having an ethical or moral dimension, which would transcend financial cost. Even so, there are some efforts underway to systemically recognize environmental costs and assets, and account for them properly in economic terms.

While affected industries spark controversy in fighting regulation, there are also many environmentalists and public interest groups who believe that current regulations are inadequate, and advocate for stronger protection. Environmental law conferences - such as the annual Public Interest Environmental Law Conference in Eugene, Oregon - typically have this focus, also connecting environmental law with class, race, and other issues.

Around the World

International Law

Global and regional environmental issues are increasingly the subject of international law. Debates over environmental concerns implicate core principles of international law and have been the subject of numerous international agreements and declarations.

Customary international law is an important source of international environmental law. These are the norms and rules that countries follow as a matter of custom and they are so prevalent that they bind all states in the world. When a principle becomes customary law is not clear cut and many arguments are put forward by states not wishing to be bound. Examples of customary international law relevant to the environment include the duty to warn other states promptly about icons of an environmental nature and environmental damages to which another state or states may be exposed, and Principle 21 of the Stockholm Declaration ('good neighbourliness' or sic utere).

Numerous legally binding international agreements encompass a wide variety of issue-areas, from terrestrial, marine and atmospheric pollution through to wildlife and biodiversity protection. International environmental agreements are generally multilateral (or sometimes bilateral) treaties (a.k.a. convention, agreement, protocol, etc.). Protocols are subsidiary agreements built from a primary treaty. They exist in many areas of international law but are especially useful in the environmental field, where they may be used to regularly incorporate recent scientific knowledge. They also permit countries to reach agreement on a framework that would be contentious if every detail were to be agreed upon in advance. The most widely known protocol in international environmental law is the Kyoto Protocol, which followed from the United Nations Framework Convention on Climate Change.

While the bodies that proposed, argued, agreed upon and ultimately adopted existing international agreements vary according to each agreement, certain conferences, including 1972's United Nations Conference on the Human Environment, 1983's World Commission on Environment and Development, 1992's United Nations Conference on Environment and Development and 2002's World Summit on Sustainable Development have been particularly important. Multilateral environmental agreements sometimes create an International Organization, Institution or Body responsible for implementing the agreement. Major examples are the Convention on International Trade in Endangered Species of Wild Fauna and Flora (CITES) and the International Union for Conservation of Nature (IUCN).

International environmental law also includes the opinions of international courts and tribunals. While there are few and they have limited authority, the decisions carry much weight with legal commentators and are quite influential on the development of international environmental law. One of the biggest challenges in international decisions is to determine an adequate compensation for environmental damages. The courts include the International Court of Justice (ICJ); the international Tribunal for the Law of the Sea (ITLOS); the European Court of Justice; European Court of Human Rights and other regional treaty tribunals.

Africa

According to the International Network for Environmental Compliance and Enforcement (INECE), the major environmental issues in Africa are "drought and flooding, air pollution, deforestation, loss of biodiversity, freshwater availability, degradation of soil and vegetation, and widespread poverty." The U.S. Environmental Protection Agency (EPA) is focused on the "growing urban and industrial pollution, water quality, electronic waste and indoor air from cookstoves." They hope to provide enough aid on concerns regarding pollution before their impacts contaminate the African environment as well as the global environment. By doing so, they intend to "protect human health, particularly vulnerable populations such as children and the poor." In order to accomplish these goals in Africa, EPA programs are focused on strengthening the ability to enforce environmental laws as well as public compliance to them. Other programs work on developing stronger environmental laws, regulations, and standards.

Asia

The Asian Environmental Compliance and Enforcement Network (AECEN) is an agreement between 16 Asian countries dedicated to improving cooperation with environmental laws in Asia. These countries include Cambodia, China, Indonesia, India, Maldives, Japan, Korea, Malaysia, Nepal, Philippines, Pakistan, Singapore, Sri Lanka, Thailand, Vietnam, and Lao PDR.

European Union

The European Union issues secondary legislation on environmental issues that are valid throughout the EU (so called regulations) and many directives that must be implemented into national legislation from the 28 member states (national states). Examples are the Regulation (EC) No. 338/97 on the implementation of CITES; or the Natura 2000 network the centerpiece for nature & biodiversity policy, encompassing the bird Directive (79/409/EEC/ changed to 2009/147/EC) and the habitats directive (92/43/EEC). Which are made up of multiple SACs (Special Areas of Conservation, linked to the habitats directive) & SPAs (Special Protected Areas, linked to the bird directive), throughout Europe.

EU legislation is ruled in Article 249 Treaty for the Functioning of the European Union (TFEU). Topics for common EU legislation are:

- Climate change
- Air pollution
- Water protection and management
- Waste management
- Soil protection
- Protection of nature, species and biodiversity
- Noise pollution
- Cooperation for the environment with third countries (other than EU member states)
- Civil protection

Middle East

The U.S. Environmental Protection Agency is working with countries in the Middle East to improve "environmental governance, water pollution and water security, clean fuels and vehicles, public participation, and pollution prevention."

Oceania

The main concerns on environmental issues in the Oceanic Region are "illegal releases of air and water pollutants, illegal logging/timber trade, illegal shipment of hazardous wastes, including e-waste and ships slated for destruction, and insufficient institutional structure/lack of enforcement capacity". The Secretariat of the Pacific Regional Environmental Programme (SPREP) is an international organization between Australia, the Cook Islands, FMS, Fiji, France, Kiribati, Marshall Islands, Nauru, New Zealand, Niue, Palau, PNG, Samoa, Solomon Island, Tonga, Tuvalu, USA, and Vanuatu. The SPREP was established in order to provide assistance in improving and protecting the environment as well as assure sustainable development for future generations.

Australia

The Environment Protection and Biodiversity Conservation Act 1999 is the center piece of environmental legislation in the Australian Government. It sets up the "legal framework to protect and manage nationally and internationally important flora, fauna, ecological communities and heritage places". It also focuses on protecting world heritage properties, national heritage properties, wetlands of international importance, nationally threatened species and ecological communities, migratory species, Commonwealth marine areas, Great Barrier Reef Marine Park, and the environment surrounding nuclear activities. *Commonwealth v Tasmania* (1983), also known as the "Tasmanian Dam Case", is the most influential case for Australian environmental law.

Brazil

The Brazilian government created the Ministry of Environment in 1992 in order to develop better strategies of protecting the environment, use natural resources sustainably, and enforce public environmental policies. The Ministry of Environment has authority over policies involving environment, water resources, preservation, and environmental programs involving the Amazon.

Canada

The Department of the Environment Act establishes the Department of the Environment in the Canadian government as well as the position Minister of the Environment. Their duties include "the preservation and enhancement of the quality of the natural environment, including water, air and soil quality; renewable resources, including migratory birds and other non-domestic flora and fauna; water; meteorology;" The Environmental Protection Act is the main piece of Canadian environmental legislation that was put into place March 31, 2000. The Act focuses on "respecting pollution prevention and the protection of the environment and human health in order to contribute to sustainable development." Other principle federal statutes include the Canadian Environmental Assessment Act, and the Species at Risk Act. When provincial and federal legislation are in conflict federal legislation takes precedence, that being said individual provinces can have their own legislation such as Ontario's Environmental Bill of Rights, and Clean Water Act.

China

According to the U.S. Environmental Protection Agency, "China has been working with great determination in recent years to develop, implement, and enforce a solid environmental law framework. Chinese officials face critical challenges in effectively implementing the laws, clarifying the roles of their national and provincial governments, and strengthening the operation of their legal system." Explosive economic and industrial growth in China has led to significant environmental degradation, and China is currently in the process of developing more stringent legal controls. The harmonization of Chinese society and the natural environment is billed as a rising policy priority.

Ecuador

With the enactment of the 2008 Constitution, Ecuador became the first country in the world to codify the Rights of Nature. The Constitution, specifically Articles 10 and 71-74, recognizes the inalienable rights of ecosystems to exist and flourish, gives people the authority to petition on the behalf of ecosystems, and requires the government to remedy violations of these rights. The rights approach is a break away from traditional environmental regulatory systems, which regard nature as property and legalize and manage degradation of the environment rather than prevent it.

The Rights of Nature articles in Ecuador's constitution are part of a reaction to a combination of political, economic, and social phenomena. Ecuador's abusive past with the oil industry, most famously the class-action litigation against Chevron, and the failure of an extraction-based economy and neoliberal reforms to bring economic prosperity to the region has resulted in the election of a New Leftist regime, led by President Rafael Correa, and sparked a demand for new approaches to development. In conjunction with this need, the principle of "Buen Vivir," or good living—focused on social, environmental and spiritual wealth versus material wealth—gained popularity among citizens and was incorporated into the new constitution.

The influence of indigenous groups, from whom the concept of "Buen Vivir" originates, in the forming of the constitutional ideals also facilitated the incorporation of the Rights of Nature as a basic tenet of their culture and conceptualization of "Buen Vivir."

Egypt

The Environmental Protection Law outlines the responsibilities of the Egyptian government to "preparation of draft legislation and decrees pertinent to environmental management, collection of data both nationally and internationally on the state of the environment, preparation of periodical reports and studies on the state of the environment, formulation of the national plan and its projects, preparation of environmental profiles for new and urban areas, and setting of standards to be used in planning for their development, and preparation of an annual report on the state of the environment to be prepared to the President."

India

In India, Environmental law is governed by the Environment Protection Act, 1986. This act is enforced by the Central Pollution Control Board and the numerous State Pollution Control Boards. Apart from this, there are also individual legislations specifically enacted for the protection of Water, Air, Wildlife, etc. Such legislations include :-

- The Water (Prevention and Control of Pollution) Act, 1974
- The Water (Prevention and Control of Pollution) Cess Act, 1977
- The Forest (Conservation) Act, 1980
- The Air (Prevention and Control of Pollution) Act, 1981
- Air (Prevention and Control of Pollution) (Union Territories) Rules, 1983
- The Biological Diversity Act, 2002 and the Wild Life Protection Act, 1972.

- Batteries (Management and Handling) Rules, 2001

- Recycled Plastics, Plastics Manufacture and Usage Rules, 1999

- The National Green Tribunal established under the National Green Tribunal Act of 2010 has jurisdiction over all environmental cases dealing with a substantial environmental question and acts covered under the Water (Prevention and Control of Pollution) Act, 1974;

- Water (Prevention and Control of Pollution) Cess Rules, 1978

- Ganga Action Plan, 1986

- The Forest (Conservation) Act, 1980

- The Public Liability Insurance Act, 1991 and the Biological Diversity Act, 2002. The acts covered under Indian Wild Life Protection Act 1972 do not fall within the jurisdiction of the National Green Tribunal. Appeals can be filed in the Hon'ble Supreme Court of India.

- Basel Convention on Control of TransboundaryMovements on Hazardous Wastes and Their Disposal, 1989 and Its Protocols

- Hazardous Wastes (Management and Handling) Amendment Rules, 2003

Japan

The Basic Environmental Law is the basic structure of Japan's environmental policies replacing the Basic Law for Environmental Pollution Control and the Nature Conservation Law. The updated law aims to address "global environmental problems, urban pollution by everyday life, loss of accessible natural environment in urban areas and degrading environmental protection capacity in forests and farmlands."

The three basic environmental principles that the Basic Environmental Law follows are "the blessings of the environment should be enjoyed by the present generation and succeeded to the future generations, a sustainable society should be created where environmental loads by human activities are minimized, and Japan should contribute actively to global environmental conservation through international cooperation." From these principles, the Japanese government have established policies such as "environmental consideration in policy formulation, establishment of the Basic Environment Plan which describes the directions of long-term environmental policy, environmental impact assessment for development projects, economic measures to encourage activities for reducing environmental load, improvement of social infrastructure such as sewerage system, transport facilities etc., promotion of environmental activities by corporations, citizens and NGOs, environmental education, and provision of information, promotion of science and technology."

New Zealand

The Ministry for the Environment and Office of the Parliamentary Commissioner for the Environment were established by the Environment Act 1986. These positions are responsible for advising the Minister on all areas of environmental legislation. A common theme of New Zealand's environmental legislation is sustainably managing natural and physical resources, fisheries, and forests. The Resource Management Act 1991 is the main piece of environmental legislation that outlines the government's strategy to managing the "environment, including air, water soil, biodiversity, the coastal environment, noise, subdivision, and land use planning in general."

Russia

The Ministry of Natural Resources and Environment of the Russian Federation makes regulation regarding "conservation of natural resources, including the subsoil, water bodies, forests located in designated conservation areas, fauna and their habitat, in the field of hunting, hydrometeorology and related areas, environmental monitoring and pollution control, including radiation monitoring and control, and functions of public environmental policy making and implementation and statutory regulation."

Vietnam

Vietnam is currently working with the U.S. Environmental Protection Agency on dioxin remediation and technical assistance in order to lower methane emissions. In March 2002, the U.S and Vietnam signed the U.S.-Vietnam Memorandum of Understanding on Research on Human Health and the Environmental Effects of Agent Orange/Dioxin.

Environmental Justice

Environmental justice emerged as a concept in the United States in the early 1980s. The term has two distinct uses. The first and more common usage describes a social movement whose focus is on the fair distribution of environmental benefits and burdens. Second, it is an interdisciplinary body of social science literature that includes theories of the environment, theories of justice, environmental law and governance, environmental policy and planning, development, sustainability, and political ecology.

Definition

The United States Environmental Protection Agency defines environmental justice as follows:

Environmental justice is the fair treatment and meaningful involvement of all people regardless of race, color, national origin, or income with respect to the development, implementation, and enforcement of environmental laws, regulations, and policies. EPA has this goal for all communities and persons across this Nation [sic]. It will be achieved when everyone enjoys the same degree of protection from environmental and health hazards and equal access to the decision-making process to have a healthy environment in which to live, learn, and work.

Other definitions include equitable distribution of environmental risks and benefits; fair and meaningful participation in environmental decision-making; recognition of community ways of life, local knowledge, and cultural difference; and the capability of communities and individuals to function and flourish in society.

Environmental Discrimination

One issue that environmental justice seeks to address is that of environmental discrimination. Racism and discrimination against minorities center on a socially-dominant group's belief in its superiority, often resulting in a) privilege for the dominant group and b) the mistreatment of non-dominant minorities. The combined impact of these privileges and prejudices are just one of the potential reasons that waste management and highly-pollutive sites tend to be located in minority-dominated areas. A disproportionate quantity of minority communities (for example in Warren County, North Carolina) play host to landfills, incinerators, and other potentially toxic facilities.

Environmental discrimination has historically been evident in the process of selecting and building environmentally hazardous sites, including waste disposal, manufacturing, and energy production facilities. The location of transportation infrastructures, including highways, ports, and airports, has also been viewed as a source of environmental injustice. Among the earliest documentation of environmental racism was a study of the distribution of toxic waste sites across the United States. Due to the results of that study, waste dumps and waste incinerators have been the target of environmental justice lawsuits and protests.

Litigation

Some environmental justice lawsuits are based on violations of civil rights laws.

Title VI of the Civil Rights Act of 1964 is often used in lawsuits that claim environmental inequality. Section 601 prohibits discrimination based on race, color, or national origin by any government agency receiving federal assistance. To win an environmental justice case that claims an agency violated this statute, the plaintiff must prove the agency intended to discriminate. Section 602 requires agencies to create rules and regulations that uphold section 601. This section is useful because the plaintiff must only prove that the rule or regulation in question had a discriminatory impact. There is no need to prove discriminatory intent. *Seif v. Chester Residents Concerned for Quality Living* set the precedent that citizens can sue under section 601. There has not yet been a case in which a citizen has sued under section 602, which calls into question whether this right of action exists.

The Equal Protection Clause of the Fourteenth Amendment, which was used many times to defend minority rights during the 1960s, has also been used in numerous environmental justice cases.

Initial Barriers to Minority Participation

When environmentalism first became popular during the first half of the 20th century, the focus was wilderness protection and wildlife preservation. These goals reflected the interests of the movement's initial supporters. The actions of many mainstream environmental organizations still reflect these early principles.

Many low-income minorities felt isolated or even negatively impacted by the movement, exemplified by the Southwest Organizing Project's (SWOP) Letter to the Group of 10, a letter sent to major environmental organizations by several local environmental justice activists. The letter argued that the environmental movement was so concerned about cleaning up and preserving nature that it ignored the negative side-effects that doing so caused communities nearby, namely less job growth. In addition, the NIMBY movement has transferred locally unwanted land uses (LULUs) from middle-class neighborhoods to poor communities with large minority populations. Therefore, vulnerable communities with fewer political opportunities are more often exposed to hazardous waste and toxins. This has resulted in the PIBBY principle, or at least the PIMBY (Place-in-minorities'-backyard), as supported by the United Church of Christ's study in 1987.

As a result, some minorities have viewed the environmental movement as elitist. Environmental elitism manifested itself in three different forms:

1. *Compositional* – Environmentalists are from the middle and upper class.

2. *Ideological* – The reforms benefit the movement's supporters but impose costs on nonparticipants.

3. *Impact* – The reforms have "regressive social impacts". They disproportionately benefit environmentalists and harm underrepresented populations.

Supporters of economic growth have taken advantage of environmentalists' neglect of minorities. They have convinced minority leaders looking to improve their communities that the economic benefits of industrial facility and the increase in the number of jobs are worth the health risks. In fact, both politicians and businesses have even threatened imminent job loss if communities do not accept hazardous industries and facilities. Although in many cases local residents do not actually receive these benefits, the argument is used to decrease resistance in the communities as well as avoid expenditures used to clean up pollutants and create safer workplace environments.

Cost Barriers

One of the major initial barriers to minority participation in environmental justice is the initial costs of trying to change the system and prevent companies from dumping their toxic waste and other pollutants in areas with high numbers of minorities living in them. There are massive legal fees involved in fighting for environmental justice and trying to shed environmental racism. For example, in the United Kingdom, there is a rule that the claimant may have to cover the fees of their opponents, which further exacerbates any cost issues, especially with lower income minority groups; also, the only way for environmental justice groups to hold companies accountable for their pollution and breaking any licensing issues over waste disposal would be to sue the government for not enforcing rules. This would lead to the forbidding legal fees that most could not afford. This can be seen by the fact that out of 210 judicial review cases between 2005 and 2009, 56% did not proceed due to costs.

Contributions of the Civil Rights Movement

During the African-American Civil Rights Movement in the 1960s, activists participated in a social movement that created a unified atmosphere and advocated goals of social justice and equality. The community organization and the social values of the era have translated to the Environmental Justice movement.

Similar Goals and Tactics

The Environmental Justice movement and the Civil Rights Movement have many commonalities. At their core, the goals of movements are the same: "social justice, equal protection, and an end to institutional discrimination." By stressing the similarities of the two movements, it emphasizes that environmental equity is a right for all citizens. Because the two movements have parallel goals, it is useful to employ similar tactics that often emerge on the grassroots level. Common confrontational strategies include protests, neighborhood demonstrations, picketing, political pressure, and demonstration.

Existing Organizations and Leaders

Just as the civil rights movement of the 1960s began in the South, the modern civil rights movement and the fight for environmental equity has been largely based in the South, where environmental discrimination is most prominent. In these southern communities, black churches and other voluntary associations are used to organize resistance efforts, including research and demonstrations, such as the protest in Warren County, North Carolina. As a result of the existing community structure, many church leaders and civil rights activists, such as Reverend Benjamin Chavis Muhammad, have spearheaded the Environmental Justice movement.

The Bronx, in New York city, has become a recent example of Environmental Justice succeeding. Majora Carter spearheaded the South Bronx Greenway Project, bringing local economic development, local urban heat island mitigation, positive social influences, access to public open space, and aesthetically stimulating environments. The New York City Department of Design and Construction has recently recognized the value of the South Bronx Greenway design, and consequently utilized it as a widely distributed smart growth template. This venture is the ideal shovel-ready project with over $50 million in funding.

Litigation

Some of the most successful Environmental Justice lawsuits are based on violations of civil rights laws. The first case to use civil rights as a means to legally challenge the siting of a waste facility was in 1979. With the legal representation of Linda McKeever Bullard, the wife of Robert D. Bullard, residents of Houston's Northwood Manor opposed the decision of the city and Browning Ferris Industries to construct a solid waste facility near their mostly African-American neighborhood.

In 1979, Northeast Community Action Group or NECAG, was formed by African American homeowners in a suburban, middle income neighborhood in order to keep a landfill out of their home town. This group was the first organization that found the connection between race and pollution. The group, alongside their attorney Linda McKeever Bullard started the lawsuit Bean v. Southwestern Waste Management, Inc., which was the first of its kind to challenge the sitting of a waste facility under civil rights law. The Equal Protection Clause of the Fourteenth Amendment, which was used many times to defend minority rights during the 1960s, has also been used in numerous Environmental Justice cases.

Title VI of the Civil Rights Act of 1964 is often used in lawsuits that claim environmental inequality. The two most important sections in these cases are sections 601 and 602. Section 601 prohibits discrimination based on race, color, or national origin by any government agency receiving federal assistance. To win an Environmental Justice case that claims an agency violated this statute, the plaintiff must prove the agency intended to discriminate. Section 602 requires agencies to create rules and regulations that uphold section 601. This section is useful because the plaintiff must only prove that the rule or regulation in question had a discriminatory impact. There is no need to prove discriminatory intent. *Seif v. Chester Residents Concerned for Quality Living* set the precedent that citizens can sue under section 601, there has not been a case in which a citizen has sued under section 602, which calls into question whether this right of action exists.

Affected Groups

Among the affected groups of Environmental Justice, those in high-poverty and racial minority groups have the most propensity to receive the harm of environmental injustice. Poor people account for more than 20% of the human health impacts from industrial toxic air releases, compared to 12.9% of the population nationwide. This does not account for the inequity found among individual minority groups. Some studies that test statistically for effects of race and ethnicity, while controlling for income and other factors, suggest racial gaps in exposure that persist across all bands of income

African-Americans are affected by a variety of Environmental Justice issues. One notorious example is the "Cancer Alley" region of Louisiana. This 85-mile stretch of the Mississippi River between Baton Rouge and New Orleans is home to 125 companies that produce one quarter of the petrochemical products manufactured in the United States. The United States Commission on Civil Rights has concluded that the African-American community has been disproportionately affected by Cancer Alley as a result of Louisiana's current state and local permit system for hazardous facilities, as well as their low socio-economic status and limited political influence.

Indigenous groups are often the victims of environmental injustices. Native Americans have suffered abuses related to uranium mining in the American West. Churchrock, New Mexico, in Navajo territory was home to the longest continuous uranium mining in any Navajo land. From 1954 until 1968, the tribe leased land to mining companies who did not obtain consent from Navajo families or report any consequences of their activities. Not only did the miners significantly deplete the limited water supply, but they also contaminated what was left of the Navajo water supply with uranium. Kerr-McGee and United Nuclear Corporation, the two largest mining companies, argued that the Federal Water Pollution Control Act did not apply to them, and maintained that Native American land is not subject to environmental protections. The courts did not force them to comply with US clean water regulations until 1980.

The most common example of environmental injustice among Latinos is the exposure to pesticides faced by farmworkers. After DDT and other chlorinated hydrocarbon pesticides were banned in the United States in 1972, farmers began using more acutely toxic organophosphate pesticides such as parathion. A large portion of farmworkers in the US are working illegally, and as a result of their political disadvantage, are not able to protest against regular exposure to pesticides. Exposure to chemical pesticides in the cotton industry also affects farmers in India and Uzbekistan. Banned throughout much of the rest of the world because of the potential threat to human health and the natural environment, Endosulfan is a highly toxic chemical, the safe use of which cannot be guaranteed in the many developing countries it is used in. Endosulfan, like DDT, is an organochlorine and persists in the environment long after it has killed the target pests, leaving a deadly legacy for people and wildlife.

Residents of cities along the US-Mexico border are also affected. Maquiladoras are assembly plants operated by American, Japanese, and other foreign countries, located along the US-Mexico border. The maquiladoras use cheap Mexican labor to assemble imported components and raw material, and then transport finished products back to the United States. Much of the waste ends up being illegally dumped in sewers, ditches, or in the desert. Along the Lower Rio Grande Valley, maquiladoras dump their toxic wastes into the river from which 95 percent of residents obtain their drinking water. In the border cities of Brownsville, Texas and Matamoros, Mexico, the rate of anencephaly (babies born without brains) is four times the national average.

One reason for toxic industries to concentrate in minority neighborhoods or poor neighborhoods is because of their lack of political power. Whether it be lack of homeownership or just because of a general inability to participate politically, these groups are treated unfairly. This lack of political participation could indicate why latinos are the most affected by environmental injustice in the US, since many latinos are illegal immigrants and thus cannot participate in the political system.

States may also see placing toxic facilities near poor neighborhoods as beneficial from a Cost Benefit Analysis (CBA) perspective. Viewing a state's wealth through the lens of CBA's, it would be more favorable to place a toxic facility near a city of 20,000 poor people than it would be to place it by a city of 5,000 wealthy people. Terry Bossert of Range Resources reportedly has said that it deliberately locates its operations in poor neighbourhoods instead of wealthy areas where residents have more money to challenge its practices.

Steel works, blast furnaces, rolling and finishing mills, along with iron and steel foundries, are responsible for more than 57% of the total human health risks from industrial pollution. This means that if the government wanted to make major reformative legislation for Environmental Justice, they could easily do so by targeting these industries.

Government Agencies

U.S. Department of Agriculture

In its 2012 environmental justice strategy documents, the U.S. Department of Agriculture (USDA) stated an ongoing desire to integrate environmental justice into its core mission, internal operations and programming. It identified ambitious timeframes for action and promised improved efforts to highlight, track and coordinate EJ activities among its many sub-agencies. Agency-wide the USDA expanded its perspective on EJ, so that in addition to preventing disproportionate environmental impacts on EJ communities, USDA voiced a commitment to improve public participation processes and use its technical and financial assistance programs to improve the quality of life in all communities. In 2011, Secretary of Agriculture Tom Vilsack emphasized the USDA's focus on EJ in rural communities around the United States. USDA funds or implements many creative programs with social and environmental equity goals, however it has no staff dedicated solely to EJ, and faces the challenges of limited budgets and coordinating the efforts of a highly diverse agency.

Background

The USDA is the executive agency responsible for federal policy on food, agriculture, natural resources, and quality of life in rural America. The USDA has more than 100,000 employees and delivers over \$96.5 billion in public services to programs worldwide. To fulfill its general mandate, USDA's departments are organized into seven mission areas:1) Farm and Foreign Agricultural Services; 2) Food, Nutrition and Consumer Services; 3) Food Safety; 4) Marketing and Regulatory Programs; 5) Natural Resources and Environment; 6) Research, Education and Economics and; 7) Rural Development.

In 1994, President Clinton issued Executive Order 12898, "Federal Actions to Address Environmental Justice in Minority Populations and Low-Income Populations." Executive Order 12898 requires that achieving EJ must be part of each federal agency's mission. Agency programs, policies and activities can lead to health and environmental effects that disproportionately impact minority and low-income populations. Under Executive Order 12898 agencies must develop strategies that identify and address these effects by:

1. promoting enforcement of all health and environmental statutes in areas with minority and low-income populations;

2. ensuring greater public participation;

3. improving research and data collection relating to the health and environment of minority and low-income populations; and

4. identifying differential patterns of consumption of natural resources among minority and low-income populations.

Title VI of the Civil Rights Act of 1964 requires that federal funds be used in a fair and equitable manner. Under Title VI any federal agency that receives federal funding cannot discriminate. Title VI also forbids federal agencies from providing grants or funding opportunities to programs that discriminate. An agency that violates Title VI can lose its federal funding.

Following E.O. 12898 and USDA's initial EJ strategic plan, USDA issued its internal Environmental Justice Department Regulation (DR 5600-002) in 1997. Although the definition of EJ was undergoing updates in 2012, DR 5600-002 defines environmental justice as "to the greatest extent practicable and permitted by law, all populations are provided the opportunity to comment before decisions are rendered on, are allowed to share in the benefits of, are not excluded from, and are not affected in a disproportionately high and adverse manner by, government programs and activities affecting human health or the environment." Patrick Holmes, Special Assistant to the Under Secretary for Natural Resources and Environment at USDA, notes that this definition will be broadened in 2012 so that EJ also includes efforts to improve quality of life in all communities. In other words, USDA will consider EJ to include avoiding adverse impacts *and* ensuring access to environmental benefits. Further, DR 5600-002 identified USDA's goals in implementing Executive Order 12898 as:

•   To incorporate environmental justice considerations into USDA's programs and activities and to address environmental justice across mission areas;

•   To identify, prevent, and/or mitigate, to the greatest extent practicable, disproportionately high and adverse human health or environmental effects of USDA programs and activities on minority and low-income populations; and

- To provide, to the greatest extent practicable, the opportunity for minority and low-income populations to participate in planning, analysis, and decisionmaking that affects their health or environment, including identification of program needs and designs.

DR 5600-002 is "intended only to improve the internal management of USDA," and although it described concrete, mandatory actions by the agency, it did not establish new rights or benefits enforceable in court. In April 2011, USDA Secretary Tom Vilsack has stated a more concrete priority to fulfill its mission of environmental justice in rural areas.

2012 Environmental Justice Strategy

In compliance with the August 2011 Memorandum of Understanding on Environmental Justice and Executive Order 12898 (MOU), USDA released a final Environmental Justice Strategic Plan: 2012 to 2014 on February 7, 2012 (Strategic Plan), which identifies new and updated goals and performance measures beyond what USDA identified in a 1995 EJ strategy it adopted in response to E.O. 12898. In the same week, it also released its first annual implementation progress report (Progress Report), as the MOU also required. The Secretary's message accompanying the Strategic Plan described two immediate tasks: 1) each agency within USDA is required to identify a point of contact for EJ issues, at the Senior Executive Service (SES) level; and 2) each agency must develop its own EJ strategy prior to April 15, 2012, and begin implementing it as soon as possible. As of May 2012, it did not appear that such strategies had been made public, although sub-agencies provided internal reports to the USDA's EJ steering committee on April 9, 2012, according to Holmes. The Secretary's message contained strong language that, "Given that USDA programs touch almost every American every day, the Department is well positioned to help in [the environmental justice] effort." USDA has determined that it can achieve the requirements of the Executive Order by integrating EJ into its programs, rather than implementing new and costly programs. The agency took this same approach in an EJ strategy it adopted in 1995. In some areas, such as agricultural chemicals and effects to migrant workers, USDA reviews its practices to identify potential disproportionate, adverse impacts on EJ communities, according to Blake Velde, Senior Environmental Scientist with the USDA Hazardous Materials Management Division. Generally, however, USDA believes its existing technical and financial assistance programs provide solutions to environmental inequity, such as its initiatives on education, food deserts, and economic development in impacted communities, and ensuring access to environmental benefits is the focus of USDA's EJ efforts.

Natural Resources and Environment (NRE) Under Secretary Harris Sherman is the political appointee generally responsible for USDA's EJ strategy, with Patrick Holmes, a senior staffer to the Under Secretary, playing a coordinating role. Although USDA has no staff dedicated solely to EJ, its sub-agencies have many offices dedicated to civil rights compliance, outreach and communication and environmental review whose responsibilities incorporate EJ issues. The Strategic Plan was developed with the input of an Environmental Justice Working Group, made up of staff and leadership representing the USDA's seven mission areas and the SES-level contacts, which were appointed in early 2012, serve as a steering committee for the agency's efforts. The Strategic Plan is organized according to six goals, which were purposefully left broad, and lists specific objectives and agency performance measures under each goal. The details and specific implementation of many of these programs and the performance measures are left to the departments and sub-agencies to develop. The six goals are to:

- Ensure USDA programs provide opportunities for EJ communities.

- Provide targeted training and capacity-building to EJ communities.

- Expand public participation in agency activities, to enhance the "credibility and public trust" of the USDA.

- Ensure USDA's activities do not have disproportionately high and adverse human health impacts, and resolve environmental justice issues and complaints.

- Increase the awareness of EJ issues among USDA employees.

- Update and/or Develop Departmental and Agency Regulations on EJ.

The Strategic Plan also lists existing programs that either currently support the goal, or are expected to in the future. According to Holmes, some of the challenges of the Strategic Plan process have stemmed from the diverse programs and missions that the agency serves, limitations on staff time, and budgets.

Environmental Justice Initiatives

The Strategic Plan requires that EJ must be integrated into the strategies and evaluations for sub-agencies' technical and financial assistance programs. It also emphasizes public participation, community capacity-building, EJ awareness and training within the USDA.

Transparency, Accountability, Accessibility and Community Participation

A stated goal of USDA's Strategic Plan is to expand public participation in agency activities, to enhance the "credibility and public trust" of the USDA. Specifically, the agency will update its public participation guidelines to include EJ, beginning this process by April 15, 2012. The Strategic Plan emphasizes capacity-building in EJ communities, and includes objectives that emphasize communication between USDA and environmental justice communities, including Tribal consultation. Sub-agencies must announce schedules for training programs in EJ communities and to develop new, preliminary outreach materials on USDA programs by April 15, 2012. An additional performance standard is to encourage EJ communities to participate in the NEPA process, an effort the Strategic Plan requires on or before February 29, 2012, although the Strategic Plan does not articulate a standard by which this could be measured. The Strategic Plan also reiterates compliance with the Executive Orders on Tribal consultation and outreach to non-proficient English speakers, and seeks more diverse representation on regional forest advisory committees. [community participation, outreach]

Generally, the USDA's process for developing the Strategic Plan demonstrates a commitment to public involvement. The USDA EJ documents are currently housed obscurely within the Departmental Management section of the USDA website, under the Hazardous Materials Management Division, although the agency plans to update its entire site in 2012 and create a more robust EJ page. The Strategic Plan was released in draft form in December 2011 for a 30-day public comment period, and responses to general types of comments received are in the Progress Report, although the comments themselves are not online. The Secretary's message accompanying the Strategic

Plan requests that organizations and individuals to continue to contact USDA with comments on the Strategic Plan and to identify USDA programs that have been the most beneficial to their communities. The agency has a dedicated email address for this purpose. Agency leadership has asked its sub-agencies to prepare responses to additional comments that have been received, and the agency will release an interim progress report, prior to winter 2013. [community participation, outreach, education]

Internal Evaluation and Training

The Strategic Plan also seeks to increase the awareness of environmental justice issues among USDA employees. The Strategic Plan does not list any existing programs in this area, but does list a series of performance measures going forward, most of which must be met by April 15, 2012. The measures include environmental justice trainings, new web pages, and potential revisions to staff manuals and handbooks. Sub-agencies began reviewing their existing training in 2012 and in their April 9, 2012 reports to the USDA EJ steering committee, sub-agencies were asked to describe their goals for enhanced EJ training. This internal, educational undertaking appears to be new in the 2012 Strategic Plan. The Strategic Plan targets Responsible Officials, meaning office and program managers, for the trainings, as well as the SES-level points of contact required by the Secretary's message. [education, study, compliance and enforcement]

The EJ Strategy tasked each sub-agency with developing its own EJ strategy document by spring 2012, although as of May 2012 the sub-agencies were still in an evaluation stage and had not issued final documents. For many sub-agencies, the 2012 process has been their first focused assessment of their EJ impact and opportunities. Going forward, sub-agencies will submit twice-yearly reports to NRE about their implementation of the Strategic Plan's goals; the first of these was due April 9, 2012, and as of May 2012, the USDA's EJ steering committee was evaluating the first reports.

Establishment of Performance Metrics

As part of its effort to ensure that EJ communities have the opportunity to participate in USDA programs, the Strategic Plan requires each sub-agency to set measurements through which it can track increased EJ community participation in USDA technical and financial assistance programs. This must be done by April 15, 2012. As of late April 2012, the sub-agencies were still in the process of describing a baseline of current activities and determining the metrics to evaluate improvement, such as staff time, grant funding or increased programming. The ultimate metrics are likely to be somewhat subjective, and must be flexible given the broad range of undertakings by the sub-agencies. Also related to evaluation, the Strategic Plan requires the sub-agencies to determine an effective methodology with which they can evaluate whether USDA programs have disproportionate impacts. [study, redressing environmental racism, compliance and enforcement]

Other Ej Initiatives

Tribal Outreach

USDA has had a role in implementing Michelle Obama's *Let's Move* campaign in Tribal Areas, by increasing participation by Bureau of Indian Education schools in Federal nutrition programs, in the development of community gardens on Tribal lands, and in the development of Tribal food policy councils. This is combined with measures to provide Rural Development funding for community infrastructure in Indian Country. [children's issues, education, diet, grants, Native Americans, public health].

The U.S. Forest Service (USFS) is working to update its policy on protection and management of Native American Sacred Sites, an effort that has included listening sessions and government-to-government consultation. The Animal and Plant Health Inspection Service (APHIS) has also consulted with Tribes regarding management of reintroduced of species, where Tribes may have a history of subsistence-level hunting of those species. Meanwhile, the Agricultural Marketing Service (AMS) is exploring a program to use meat from bisons raised on Tribal land to supply AMS food distribution programs to Tribes. [Native Americans, diet, subsistence, community participation]

The Intertribal Technical Assistance Network works to improve access of Tribal governments, communities and individuals to USDA technical assistance programs.

Technical and Financial Assistance to Farmers

The Progress Report highlights the NRCS Strike Force Initiative, which has identified impoverished counties in Mississippi, Georgia and Arkansas to receive increased outreach and training regarding USDA assistance programs. USDA credits this increased outreach with generating a 196 percent increase in contracts, representing more than 250,000 acres of farmland, in its Environmental Quality Incentives Program. [economic benefit, equitable development, grants, outreach, ej as evaluation criteria] NRCS works with "private landowners protect their natural resources" through conservation planning and assistance with the goal of maintaining "productive lands and healthy ecosystems." NRCS has its own civil rights compliance guidance document, and in 2001 NRCS funded and published a study, "Environmental Justice: Perceptions of Issues, Awareness and Assistance," focused on rural, Southern "Black Belt" counties and analyzing how the NRCS workforce could more effectively integrate environmental justice into impacted communities. [compliance and enforcement, redressing environmental racism, grants, study, ej as evaluation criteria]

The Farm Services Agency in 2011 devoted $100,000 of its Socially Disadvantaged Farmers and Ranchers program budget to improving its outreach to counties with persistent poverty, including improving its materials and building relationships with local universities and community groups. [economic benefit, equitable development, grants, outreach, ej as evaluation criteria]

In addition, USDA's Risk Management Agency has initiated education and outreach to low-income farmers regarding use of biological controls, rather than pesticides, for pest control, efforts that the agency believes are valuable in the face of climate change. [climate change, agricultural chemicals, education]

Green Jobs and Capacity Building

A 2011 MOU between a USDA sub-agency, the Food Safety Inspection Service (FSIS) and the American Indian Science and Engineering Society that aims to increase the number of Native Americans entering the FSIS career path; [education, community participation, economic benefit, green jobs, Native Americans, diet, interagency collaboration]

A partnership between APHIS and the Rural Coalition (Coalicion)--an alliance of regionally and culturally diverse organizations working to build a more just and sustainable food system. The partnership focuses on outreach, fair returns to minority and other small farmers and rural communities, farmworker working conditions, environmental protection and food safety. [agricultural chemicals, community participation, diet, economic benefit, outreach, improving health and safety, ej as evaluation criteria]

USFS is also funding pilot initiatives, such as its Urban Water Ambassadors, summer internship positions for youth who coordinate and implement urban tree planting projects. In 2011, USFS provided a grant to the Maryland Department of Natural Resources that funded 14 summer jobs for youth in Baltimore to work on urban watershed restoration programs. [community participation, green jobs, mapping, water]

Mapping

USFS has established several Urban Field Stations, to research urban natural resources' structure, function, stewardship, and benefits. By mapping urban tree coverage, the agency hopes to identify and prioritize EJ communities for urban forest projects. [community education, mapping, diet, improving health and safety, ej as evaluation criteria]

Another initiative highlighted by the agency is the Food and Nutrition Service and Economic Research Service's Food Desert Locator. The Locator provides a spatial view of food deserts, defined as a low-income census tract where a substantial number or share of residents has low access to a supermarket or large grocery store. It also shows, by census tract, the number and percentage of certain populations, such as children, seniors, or households without a vehicle, with low access to grocery stores. The mapped deserts can be used to direct agency resources to increase access to fresh fruits and vegetables and other food assistance programs, according to Blake Velde, an agency scientist and spokesperson on EJ issues. [diet, mapping, improving health and safety, study, ej as evaluation criteria, services and data available to others]

Rural Outreach

USDA Secretary Tom Vilsack has placed a clear emphasis on supporting EJ in rural areas. Although "often the highest profile battles on [environmental justice] issue[s] are waged in at-risk neighborhoods in major cities or at Superfund sites located near populated urban and suburban areas" Vilsack highlighted the often overlooked rural areas where environmental justice is largely ignored.

Through its Rural Utilities Service, the USDA supports a number of Water and Environmental Programs. These programs work to administer water and wastewater loans or grants to rural areas and cities to support water and wastewater, stormwater and solid waste disposal systems, including SEARCH grants that are targeted to financially distressed, small rural communities and other opportunities specifically for Alaskan Native villages and designated Colonias.; In his speech, Secretary Vilsack said that the USDA funded 2,575 clean water projects in rural areas during a two-year period to address problems ranging from wastewater treatment to sewage treatment. [water, land use, compliance and enforcement, improving health and safety, pollution cleanup, ej as evaluation criteria]

The USDA also supports the Rural Energy for America Grant Program. This program provides grants and loans to farmers, ranchers and rural small businesses to finance renewable energy systems and energy efficiency improvements.[grants, economic benefit, ej as evaluation criteria]

Regulations or Formalized EJ Guidelines

In 1997 the USDA promulgated a departmental regulation providing "direction to [sub-]agencies for integrating environmental justice considerations into USDA programs and activities" (DR 5600-002). Issuance of this regulation was a primary goal of USDA's 1995 EJ strategy document. DR 5600-002 includes guidelines for consideration of EJ in the NEPA process, but also stated that "efforts to address environmental justice are not limited to NEPA compliance." It requires evaluation of activities for potential disproportionate EJ impacts, outreach, and performance-metric based evaluation and reporting on sub-agencies' implementation of EJ goals. DR 5600-002 is a forward-looking, permanent directive that applies to all USDA programs and activities. However, it was not published in the Federal Register as a formal rulemaking and does not create a private right of action or enforcement tool. A Strategic Plan goal is to update this regulation, as well as other departmental regulations and policies on EJ. According to USDA, the EJ definition in DR 5600-002 will be modified in 2012—EJ to include measures to avoid disproportionate negative impacts as well as quality-of-life improvements that the agency believes can benefit impacted communities.

The Strategic Plan also has established a performance standard requiring that existing and new USDA regulations are evaluated for EJ impacts or benefits. Sub-agencies are required to develop a process for this evaluation by April 15, 2012. This performance standard reflects a requirement in DR 5600-002 that required the USDA departmental regulation on rulemaking, DR 1521-1, to be revised to require an EJ evaluation in the rulemaking process. As of 2012, DR 1521-1 requires that a cost-benefit analysis of major human health, safety and environmental regulations include analysis of risks to "persons who are disproportionately exposed or particularly sensitive," although DR 1521-1 does not mention EJ or impacts to minority or low-income communities explicitly. [Land Use - permitting, community participation, compliance and enforcement, study]

Enforcement

The Strategic Plan sets an enforcement-specific goal, which includes objectives to "effectively resolve or adjudicate all environmental justice-related Title VI complaints" and to include environmental justice as a key component of civil rights compliance reviews. Agencies are also required to identify an assessment methodology by April 15, 2012, which can be used to determine whether programs have disproportionately high and adverse environmental and human health impacts. The NRCS has published and updated a Civil Rights Compliance Review Guide, which guides the NRCS Civil Rights Division's review of the compliance with Title VI and 12898 in the agency's state offices, field offices and other facilities. The guide was updated in November 2011 and it does not mention EJ explicitly. However, the Strategic Plan identifies the NRCS compliance review and other outreach and research programs as supporting its EJ enforcement goals.

NEPA

The 1997 Regulation, DR 5600-2 required USDA sub-agencies to develop their own NEPA environmental justice guidance documents. The sub-agencies have done so, with some additional details, such as a reminder that the EJ community should be involved in identifying the alternatives, suggested stakeholders and resources, and guidance to hold meetings at times when working people can get to them, and to translate notices. However, when DR 5600-02 is updated as required by the Strategic Plan, changes could be made to the NEPA section of the Regulation. The Strategic Plan sets a performance standard to encourage interested environmental justice communities to be involved in the public participation process for NEPA documents, although the Strategic Plan does not require updates to the NEPA portions of DR 5600-02.

Although the USDA has integrated EJ into each step of the NEPA process as required by Executive Order 12898, many of the NEPA documents completed by the USDA include only cursory analysis of environmental justice effects. This analysis most often includes a rote paragraph as to what Executive Order 12898 requires and a quick conclusion that the agency action does not affect minority and low-income populations. Some examples where the USDA included more in-depth analysis are:

- Descriptions of the minority and low-income populations that live in the study area;

- Impacts relevant to socio-economic environment including changes in employment and income variations in the distribution of social welfare. [community participation, education, outreach, ej as evaluation criteria]

Permitting

The USDA does not have any permitting initiatives specific to EJ.

Title VI

The USDA has an Office of the Assistant Secretary for Civil Rights whose mission it is to provide leadership and direction "for the fair and equitable treatment of all USDA customers."

In 2003 the USDA revised DR 4300-4, internal regulations requiring a Civil Rights Impact Analysis of all "policies, actions or decisions" affecting the USDA's federally conducted and federally assisted programs or activities. The analysis is used to determine the "scope, intensity, direction, duration, and significance of the effects of an agency's proposed . . . policies, actions or decisions." USDA's departmental regulation on EJ, DR 5600-002, required DR 4300-4 to be revised to "require that Civil Rights Impact Analyses include a finding as to whether proposed or new actions have or do not have a disproportionately high and adverse effect on the human health or the environment of minority populations, and whether such effects can be prevented or mitigated." Although DR 4300-4 was revised in 2003, the revised regulation does not explicitly require a finding on adverse environmental or health impacts. [study, compliance and enforcement]

Right-to-Know Movement

A new movement, bent on educating the people, was born after the Bhopal disaster, called the "right-to-know" movement. A series of laws and reports was created, all built to inform the people of the pollutants being dumped into our neighborhoods and atmosphere, and exactly how much of each chemical is being exposed and dumped. The theory behind "right-to-know" is that once people are informed on what is polluting their neighborhood, then they will begin to take action in both bringing down their own emissions, as well as begin to make the companies causing the most pollution, through means such as protests, to take into account their actions.

Emergency Planning and Right to Know Act of 1986

After the Bhopal disaster, where a Union Carbide plant released forty tons of methyl isocyanate into the atmosphere in a village just south of Bhopal, India, the U.S. government passed the Emergency Planning and Right to Know Act of 1986. Introduced by Henry Waxman, the act required all corporations to report their toxic chemical pollution annually, which was then gathered into a report known as the Toxics Release Inventory (TRI). By collecting this data, the government was able to make sure that companies were no longer releasing excessive amounts of deadly toxins into populated areas, so to prevent another incident like that of the thousands of people killed and the tens of thousands of people injured in the Bhopal disaster.

Corporate Toxics Information Report

The Corporate Toxics Information Project (CTIP) was founded on the guidelines that they will "[develop] and [disseminate] information and analysis on corporate releases of pollutants and the consequences for communities." The overarching goal was to help take corporations into account for their pollution habits, by collecting information and putting it in databases so to make it available to the general public. The four goals of the project were to develop 1) corporate rankings, 2) regional reports, based on state, region, and metropolitan areas, 3) industry reports, based on industrial sectors, and 4) to create a web-based resource open to the entire population, that can depict all the collected data. The data collection would be done by the Environmental Protection Agency (EPA) and then analyzed and disseminated by the PERI institute.

One of the biggest projects of CTIP was the Toxic 100. The Toxic 100 is an index of the top 100 air polluters around the United States in terms of the country's largest corporations. The list is based on the EPA's Risk Screening Environmental Indicators (RSEI), which "assesses the chronic human health risk from industrial toxic releases", as well as the Toxics Release Inventory (TRI), which is where the corporations must report their chemical releases to the US government. Since its original publishing date in 2004, the Toxic 100 has been updated four more times, with the latest publishing date being August 2013.

Around the World

In recent years Environmental Justice campaigns have also emerged in other parts of the world, such as India, South Africa, Israel, Nigeria, Mexico, Hungary, Uganda, and the United Kingdom. In Europe for example, there is evidence to suggest that the Romani people and other minority groups of non-European descent are suffering from environmental inequality and discrimination.

In Europe

In Europe, the Romani peoples are ethnic minorities and differ from the rest of the European people by their culture, language, and history. The environmental discrimination that they experience ranges from the unequal distribution of environmental harms as well as the unequal distribution of education, health services and employment. In many countries Romani peoples are forced to live in the slums because many of the laws to get residence permits are discriminatory against them. This forces Romani people to live in urban "ghetto" type housing or in shantytowns. In the Czech Republic and Romania, the Romani peoples are forced to live in places that have less access to running water and sewage, and in Ostrava, Czech Republic, the Romani people live in apartments located above an abandoned mine, which emits methane. Also in Bulgaria, the public infrastructure extends throughout the town of Sofia until it reaches the Romani village where there is very little water access or sewage capacity.

The European Union is trying to strive towards environmental justice by putting into effect declarations that state that all people have a right to a healthy environment. The Stockholm Declaration, the 1987 Brundtland Commission's Report – "Our Common Future", the Rio Declaration, and Article 37 of the Charter of Fundamental Rights of the European Union, all are ways that the Europeans have put acts in place to work toward environmental justice. Europe also funds action-oriented projects that work on furthering Environmental Justice throughout the world. For example, EJOLT (Environmental Justice Organisations, Liabilities and Trade) is a large multinational project supported through the FP7 Science in Society budget line from the European Commission. From March 2011 to March 2015, 23 civil society organizations and universities from 20 countries in Europe, Africa, Latin-America, and Asia are, and have promised to work together on advancing the cause of Environmental Justice. EJOLT is building up case studies, linking organisations worldwide, and making an interactive global map of Environmental Justice.

In the United Kingdom

Whilst the predominant agenda of the Environmental Justice movement in the United States has been tackling issues of race, inequality, and the environment, environmental justice campaigns around the world have developed and shifted in focus. For example, the EJ movement in the United Kingdom is quite different. It focuses on issues of poverty and the environment, but also tackles issues of health inequalities and social exclusion. A UK-based NGO, named the Environmental Justice Foundation, has sought to make a direct link between the need for environmental security and the defense of basic human rights. They have launched several high profile campaigns that link environmental problems and social injustices. A campaign against illegal, unreported and unregulated (IUU) fishing highlighted how 'pirate' fisherman are stealing food from local, artisanal fishing communities. They have also launched a campaign exposing the environmental and human rights abuses involved in cotton production in Uzbekistan. Cotton produced in Uzbekistan is often harvested by children for little or no pay. In addition, the mismanagement of water resources for crop irrigation has led to the near eradication of the Aral Sea. The Environmental Justice Foundation has successfully petitioned large retailers such as Wal-mart and Tesco to stop selling Uzbek cotton.

Building of Alternatives to Climate Change

In France, numerous Alternatiba events, or villages of alternatives, are providing hundreds of alternatives to climate change and lack of environmental justice, both in order to raise people's awareness and to stimulate behaviour change. They have been or will be organized in over sixty different French and European cities, such as Bilbao, Brussels, Geneva, Lyon or Paris.

In South Africa

Under colonial and apartheid governments in South Africa, thousands of black South Africans were removed from their ancestral lands to make way for game parks. Earthlife Africa was formed in 1988 (www.earthlife.org.za), making it Africa's first environmental justice organisation. In 1992, the Environmental Justice Networking Forum (EJNF), a nationwide umbrella organization designed to coordinate the activities of environmental activists and organizations interested in social and environmental justice, was created. By 1995, the network expanded to include 150 member organizations and by 2000, it included over 600 member organizations.

With the election of the African National Congress (ANC) in 1994, the environmental justice movement gained an ally in government. The ANC noted "poverty and environmental degradation have been closely linked" in South Africa. The ANC made it clear that environmental inequalities and injustices would be addressed as part of the party's post-apartheid reconstruction and development mandate. The new South African Constitution, finalized in 1996, includes a Bill of Rights that grants South Africans the right to an "environment that is not harmful to their health or well-being" and "to have the environment protected, for the benefit of present and future generations through reasonable legislative and other measures that

1. prevent pollution and ecological degradation;

2. promote conservation; and

3. secure ecologically sustainable development and use of natural resources while promoting justifiable economic and social development".

South Africa's mining industry is the largest single producer of solid waste, accounting for about two-thirds of the total waste stream. Tens of thousands of deaths have occurred among mine workers as a result of accidents over the last century. There have been several deaths and debilitating diseases from work-related illnesses like asbestosis. For those who live next to a mine, the quality of air and water is poor. Noise, dust, and dangerous equipment and vehicles can be threats to the safety of those who live next to a mine as well. These communities are often poor and black and have little choice over the placement of a mine near their homes. The National Party introduced a new Minerals Act that began to address environmental considerations by recognizing the health and safety concerns of workers and the need for land rehabilitation during and after mining operations. In 1993, the Act was amended to require each new mine to have an Environmental Management Program Report (EMPR) prepared before breaking ground. These EMPRs were intended to force mining companies to outline all the possible environmental impacts of the particular mining operation and to make provision for environmental management.

In October 1998, the Department of Minerals and Energy released a White Paper entitled *A Minerals and Mining Policy for South Africa*, which included a section on Environmental Management. The White Paper states "Government, in recognition of the responsibility of the State as custodian of the nation's natural resources, will ensure that the essential development of the country's mineral resources will take place within a framework of sustainable development and in accordance with national environmental policy, norms, and standards". It adds that any environmental policy "must ensure a cost-effective and competitive mining industry."

In Australia

In Australia, the "Environmental Justice Movement" is not defined as it is in the United States. Australia does have some discrimination mainly in the siting of hazardous waste facilities in areas where the people are not given proper information about the company. The injustice that takes place in Australia is defined as environmental politics on who get the unwanted waste site or who has control over where factory opens up. The movement towards equal environmental politics focuses more on who can fight for companies to build, and takes place in the parliament; whereas, in the United States Environmental Justice is trying to make nature safer for all people.

In Ecuador

An example of the environmental injustices that indigenous groups face can be seen in the Chevron-Texaco incident in the Amazon rainforest. Texaco, which is now Chevron, found oil in Ecuador in 1964 and built sub-standard oil wells to cut costs. The deliberately used inferior technology to make their operations cheaper, even if detrimental to the local people and environment. After the company left in 1992, they left approximately one thousand toxic waste pits open and dumped billions of gallons of toxic water into the rivers.

In South Korea

South Korea has a relatively short history of environmental justice compared to other countries in the west. As a result of rapid industrialization, people started to have awareness on pollution, and from the environmental discourses the idea of environmental justice appeared. The concept of environmental justice appeared in South Korea in late 1980s.

South Korea experienced rapid economic growth (which is commonly referred to as the 'Miracle on the Han River') in the 20th century as a result of industrialization policies adapted by Park Chung-hee after 1970s. The policies and social environment had no room for environmental discussions, which aggravated the pollution in the country.

Environmental movements in South Korea started from air pollution campaigns. As the notion of environment pollution spread, the focus on environmental activism shifted from existing pollution to preventing future pollution, and the organizations eventually started to criticize the government policies that are neglecting the environmental issues. The concept of environmental justice was introduced in South Korea among the discussions of environment after 1990s. While the environmental organizations analyzed the condition of pollution in South Korea, they noticed that the environmental problems were inequitably focused especially on regions where people with low social and economic status were concentrated.

The problems of environmental injustice have arisen by environment related organizations, but approaches to solve the problems were greatly supported by the government, which developed various policies and launched institution. These actions helped raise awareness of environmental justice in South Korea. Existing environment policies were modified to cover environmental justice issues.

Environmental justice began to be widely recognized in the 1990s through policy making and researches of related institutions. For example, the Ministry of Environment, which was founded in 1992, launched Citizen's Movement for Environmental Justice (CMEJ) to raise awareness of the problem and figure out appropriate plans. As a part of its activities, Citizen's Movement for Environmental Justice (CMEJ) held Environmental Justice forum in 1999, to gather and analyze the existing studies on the issue which were done sporadically by various organizations. Citizen's Movement for Environmental Justice (CMEJ) started as a small organization, but it is keep growing and expanding. In 2002, CMEJ had more than 5 times the numbers of members and 3 times the budget it had in the beginning year.

Environmental injustice is still an ongoing problem. One example is the construction of Saemangeum Seawall. The construction of Saemangeum Seawall, which is the world's longest dyke (33 kilometers) runs between Yellow Sea and Saemangeum estuary, was part of a government project initiated in 1991. The project raised concerns on the destruction of ecosystem and taking away the local residential regions. It caught the attention of environmental justice activists because the main victims were low-income fishing population and their future generations. This is considered as an example of environmental injustice which was caused by the execution of exclusive development-centered policy.

The construction of Seoul-Incheon canal also raised environmental justice controversies. The construction took away the residential regions and farming areas of the local residents. Also, the environment worsened in the area because of the appearance of wet fogs which was caused by water deprivation and local climate changes caused by the construction of canal. The local residents, mostly people with weak economic basis, were severely affected by the construction and became the main victims of such environmental damages. While the socially and economically weak citizens suffered from the environmental changes, most of the benefits went to the industries and conglomerates with political power.

Construction of industrial complex was also criticized in the context of environmental justice. The conflict in wicheon region is one example. The region became the center of controversy when the government decided to build industrial complex of dye houses, which were formerly located in Daegu metropolitan region. As a result of the construction, Nakdong River, which is one of the main rivers in South Korea, were contaminated and local residents suffered from environmental changes caused by the construction.

Environmental justice is a growing issue in South Korea. Although the issue is not yet widely recognized compared to other countries, many organizations beginning to recognize the issue.

Between Northern and Southern Countries

Environmental discrimination in a global perspective is also an important factor when examining the Environmental Justice movement. Even though the Environmental Justice movement began in the United States, the United States also contributes to expanding the amount of environmental injustice that takes place in less-developed countries. Some companies in the United States and in other developed nations around the world contribute to the injustice by shipping the toxic waste and byproducts of factories to less-developed countries for disposal. This act increases the amount of waste in the third world countries, most of which do not have proper sanitation for their own waste much less the waste of another country. Often, the people of the less-developed countries are exposed to toxins from this waste and do not even realize what kind of waste they are encountering or the health problems that could come with it.

One prominent example of northern countries shipping their waste to southern countries took place in Haiti. Philadelphia, Pennsylvania had ash from the incineration of toxic waste that they did not have room to dump. Philadelphia decided to put the ash into the hands of a private company, which shipped the ash and dumped it in various other parts of the world, outside of the United States. *The Khian Sea*, the ship the ash was put on, sailed around the world and many countries would not accept the waste because it was hazardous for the environment and the people. The ship owners finally dumped the waste, labeled Fertilizer, in Haiti, on the beach, and sailed away in the night. The government of Haiti was infuriated and called for the waste to be removed, but the company would not come to take the ash away. The fighting over who was responsible for the waste and who would remove the waste went on for many years. After debating for over ten years, the waste was removed and taken back to a site just outside Philadelphia to be disposed of permanently.

The reason that this transporting of waste from Northern countries to the Southern countries takes place is because it is cheaper to transport waste to another country and dump it there, than to pay to dump the waste in the producing country because the third world countries do not have the same strict industry regulations as the more developed countries. The countries that the waste is taken to are usually empoverished and the governments have little or no control over the happenings in the country or do not care about the people.

Transnational Movement Networks

Many of the Environmental Justice Networks that began in the United States expanded their horizons to include many other countries and became Transnational Networks for Environmental Justice. These networks work to bring Environmental Justice to all parts of the world and protect all citizens of the world to reduce the environmental injustice happening all over the world. Listed below are some of the major Transnational Social Movement Organizations.

- Basel Action Network — works to end toxic waste dumping in poor undeveloped countries from the rich developed countries.

- GAIA (Global Anti-Incinerator Alliance) — works to find different ways to dispose of waste other than incineration. This company has people working in over 77 countries throughout the world.

- GR (Global Response) — works to educate activists and the upper working class how to protect human rights and the ecosystem.

- Greenpeace International — which was the first organization to become the global name of Environmental Justice. Greenpeace works to raise the global consciousness of transnational trade of toxic waste.

- Health Care without Harm — works to improve the public health by reducing the environmental impacts of the health care industry.

- International Campaign for Responsible Technology — works to promote corporate and government accountability with electronics and how the disposal of technology affect the environment.

- International POPs Elimination Network — works to reduce and eventually end the use of persistent organic pollutants (POPs) which are harmful to the environment.

- PAN (Pesticide Action Network) — works to replace the use of hazardous pesticides with alternatives that are safe for the environment.

These global networks work together to achieve the shared goal of a cleaner environment.

Environmental Governance

Environmental governance is a concept in political ecology and environmental policy that advocates sustainability (sustainable development) as the supreme consideration for managing all human activities—political, social and economic. Governance includes government, business and civil society, and emphasizes whole system management. To capture this diverse range of elements, environmental governance often employs alternative systems of governance, for example watershed-based management.

It views natural resources and the environment as global public goods, belonging to the category of goods that are not diminished when they are shared. This means that everyone benefits from for example, a breathable atmosphere, stable climate and stable biodiversity.

Public goods are non-rivalrous—a natural resource enjoyed by one person can still be enjoyed by others—and non-excludable—it is impossible to prevent someone consuming the good (breathing). Nevertheless, public goods are recognized as beneficial and therefore have value. The notion of a global public good thus emerges, with a slight distinction: it covers necessities that must not be destroyed by one person or state.

The non-rivalrous character of such goods calls for a management approach that restricts public and private actors from damaging them. One approach is to attribute an economic value to the resource. Water is possibly the best example of this type of good.

As of 2013 environmental governance is far from meeting these imperatives. "Despite a great awareness of environmental questions from developed and developing countries, there is environmental degradation and the appearance of new environmental problems. This situation is caused by the parlous state of global environmental governance, wherein current global environmental governance is unable to address environmental issues due to many factors. These include fragmented governance within the United Nations, lack of involvement from financial institutions, proliferation of environmental agreements often in conflict with trade measures; all these various problems disturb the proper functioning of global environmental governance. Moreover, divisions among northern countries and the persistent gap between developed and developing countries also have to be taken into account to comprehend the institutional failures of the current global environmental governance."

Definitions

What is Environmental Governance?

Environmental governance refers to the processes of decision-making involved in the control and management of the environment and natural resources. International Union for Conservation of

Nature (IUCN), define Environmental Governance as the 'Multi-level interactions (i.e., local, national, international/global) among, but not limited to, three main actors, i.e., state, market, and civil society, which interact with one another, whether in formal and informal ways; in formulating and implementing policies in response to environment-related demands and inputs from the society; bound by rules, procedures, processes, and widely accepted behavior; possessing characteristics of "good governance"; for the purpose of attaining environmentally-sustainable development' (ICUN 2014)

Key principles of environmental governance include:

- Embedding the environment in all levels of decision-making and action

- Conceptualizing cities and communities, economic and political life as a subset of the environment

- Emphasizing the connection of people to the ecosystems in which they live

- Promoting the transition from open-loop/cradle-to-grave systems (like garbage disposal with no recycling) to closed-loop/cradle-to-cradle systems (like permaculture and zero waste strategies).

Neoliberal Environmental Governance – is an approach to the theory of environmental governance framed by a perspective on neoliberalism as an ideology, policy and practice in relation to the biophysical world. There are many definitions and applications of neoliberalism, e.g. in economic, international relations, etc. However, the traditional understanding of neoliberalism is often simplified to the notion of the primacy of market-led economics through the rolling back of the state, deregulation and privatisation. Neoliberalism has evolved particularly over the last 40 years with many scholars leaving their ideological footprint on the neoliberal map. Hayek and Friedman believed in the superiority of the free market over state intervention. As long as the market was allowed to act freely, the supply/demand law would ensure the 'optimal' price and reward. In Karl Polanyi's opposing view this would also create a state of tension in which self-regulating free markets disrupt and alter social interactions and "displace other valued means of living and working". However, in contrast to the notion of an unregulated market economy there has also been a "paradoxical increase in [state] intervention" in the choice of economic, legislative and social policy reforms, which are pursued by the state to preserve the neoliberal order. This contradictory process is described by Peck and Tickell as roll back/roll out neoliberalism in which on one hand the state willingly gives up the control over resources and responsibility for social provision while on the other, it engages in "purposeful construction and consolidation of neoliberalised state forms, modes of governance, and regulatory relations".

There has been a growing interest in the effects of neoliberalism on the politics of the non-human world of environmental governance. Neoliberalism is seen to be more than a homogenous and monolithic 'thing' with a clear end point. It is a series of path-dependent, spatially and temporally "connected neoliberalisation" processes which affect and are affected by nature and environment that "cover a remarkable array of places, regions and countries". Co-opting neoliberal ideas of the importance of private property and the protection of individual (investor) rights, into environmental governance can be seen in the example of recent multilateral trade agreements (see in particular the North American Free Trade Agreement). Such neoliberal structures further reinforce a process of nature enclosure and primitive accumulation or "accumulation by dispossession" that serves to privatise increasing areas of nature. The ownership-transfer of resources traditionally not privately owned to free market mechanisms is believed to deliver greater efficiency and optimal return on investment. Other similar examples of neo-liberal inspired projects include the enclosure of minerals, the fisheries quota system in the North Pacific and the privatisation of water supply and sewage treatment in England and Wales. All three examples share neoliberal characteristics to "deploy markets as the solution to environmental problems" in which scarce natural resources are commercialized and turned into commodities. The approach to frame the ecosystem in the context of a price-able commodity is also present in the work of neoliberal geographers who subject nature to price and supply/demand mechanisms where the earth is considered to be a quantifiable resource (Costanza, for example, estimates the earth ecosystem's service value to be between 16 and 54 trillion dollars per year).

Environmental Issues

Main Drivers of Environmental Degradation

Economic growth – The development-centric vision that prevails in most countries and international institutions advocates a headlong rush towards more economic growth. Environmental economists on the other hand, point to a close correlation between economic growth and environmental degradation, arguing for qualitative development as an alternative to growth. As a result, the past couple of decades has seen a big shift towards sustainable development as an alternative to neo-liberal economics. There are those, particularly within the alternative globalization movement, who maintain that it is feasible to change to a degrowth phase without losing social efficiency or lowering the quality of life.

Consumption – The growth of consumption and the cult of consumption, or consumerist ideology, is the major cause of economic growth. Overdevelopment, seen as the only alternative to poverty, has become an end in itself. The means for curbing this growth are not equal to the task, since the phenomenon is not confined to a growing middle class in developing countries, but also concerns the development of irresponsible lifestyles, particularly in northern countries, such as the increase in the size and number of homes and cars per person.

Destruction of biodiversity – The complexity of the planet's ecosystems means that the loss of any species has unexpected consequences. The stronger the impact on biodiversity, the stronger the likelihood of a chain reaction with unpredictable negative effects. Another important factor of environmental degradation that falls under this destruction of biodiversity, and must not be ignored is deforestation. Despite all the damage inflicted, a number of ecosystems have proved to be resilient. Environmentalists are endorsing a precautionary principle whereby all potentially damaging activities would have to be analyzed for their environmental impact.

Population growth – Forecasts predict 8.9 billion people on the planet in 2050. This is a subject which primarily affects developing countries, but also concerns northern countries; although their demographic growth is lower, the environmental impact per person is far higher in these countries. Demographic growth needs to be countered by developing education and family planning programs and generally improving women's status.

"Pollution" - Pollution caused by the use of fossil fuels is another driver of environmental destruction. The burning of carbon-based fossil fuels such as coal and oil, releases carbon dioxide into the atmosphere. One of the major impacts of this is the climate change that is currently taking place on the planet, where the earth's temperature is gradually rising. Given that fuels such as coal and oil are the most heavily used fuels, this a great concern to many environmentalists.

"Agricultural practices" - Destructive agricultural practices such as overuse of fertilizers and overgrazing lead to land degradation. The soil gets eroded, and leads to silting in rivers and reservoirs. Soil erosion is a continuous cycle and ultimately results in desertification of the land. Apart from land degradation, water pollution is also a possibility; chemicals used in farming can run-off into rivers and contaminate the water.

Challenges

The crisis by the impact of human activities on nature calls for governance. Which includes responses by international institutions, governments and citizens, who should meet this crisis by pooling the experience and knowledge of each of the agents and institutions concerned.

The environmental protection measures taken remain insufficient. The necessary reforms require time, energy, money and diplomatic negotiations. The situation has not generated a unanimous response. Persistent divisions slow progress towards global environmental governance.

The global nature of the crisis limits the effects of national or sectoral measures. Cooperation is necessary between actors and institutions in international trade, sustainable development and peace.

Global, continental, national and local governments have employed a variety of approaches to environmental governance. Substantial positive and negative spillovers limit the ability of any single jurisdiction to resolve issues.

Challenges facing environmental governance include:

- Inadequate continental and global agreements
- Unresolved tensions between maximum development, sustainable development and maximum protection, limiting funding, damaging links with the economy and limiting application of Multilateral Environment Agreements (MEAs).
- Environmental funding is not self-sustaining, diverting resources from problem-solving into funding battles.
- Lack of integration of sector policies
- Inadequate institutional capacities
- Ill-defined priorities
- Unclear objectives
- Lack of coordination within the UN, governments, the private sector and civil society
- Lack of shared vision

- Interdependencies among development/sustainable economic growth, trade, agriculture, health, peace and security.

- International imbalance between environmental governance and trade and finance programs, e.g., World Trade Organization (WTO).

- Limited credit for organizations running projects within the Global Environment Facility (GEF)

- Linking UNEP, United Nations Development Programme (UNDP) and the World Bank with MEAs

- Lack of government capacity to satisfy MEA obligations

- Absence of the gender perspective and equity in environmental governance

- Inability to influence public opinion

- Time lag between human action and environmental effect, sometimes as long as a generation

- Environmental problems being embedded in very complex systems, of which our understanding is still quite weak

All of these challenges have implications on governance, however international environmental governance is necessary. The IDDRI claims that rejection of multilateralism in the name of efficiency and protection of national interests conflicts with the promotion of international law and the concept of global public goods. Others cite the complex nature of environmental problems.

tBäckstrand and Saward wrote, "sustainability and environmental protection is an arena in which innovative experiments with new hybrid, plurilateral forms of governance, along with the incorporation of a transnational civil society spanning the public-private divide, are taking place."

Local Governance

A 1997 report observed a global consensus that sustainable development implementation should be based on local level solutions and initiatives designed with and by the local communities. Community participation and partnership along with the decentralisation of government power to local communities are important aspects of environmental governance at the local level. Initiatives such as these are integral divergence from earlier environmental governance approaches which was "driven by state agendas and resource control" and followed a top-down or trickle down approach rather than the bottom up approach that local level governance encompasses. Local level governance shifts decision making power away from the state and/or governments to the grassroots. Local level governance is extremely important even on a global scale. Environmental governance at the global level is defined as international and as such has resulted in the marginalisation of local voices. Local level governance is important to bring back power to local communities in the global fight against environmental degridation. Pulgar Vidal observed a "new institutional framework, [wherein] decision-making regarding access to and use of natural resources has become increasingly decentralized." He noted four techniques that can be used to develop these processes:

- formal and informal regulations, procedures and processes, such as consultations and participative democracy;

- social interaction that can arise from participation in development programs or from the reaction to perceived injustice;

- regulating social behaviours to reclassify an individual question as a public matter;

- within-group participation in decision-making and relations with external actors.

He found that the key conditions for developing decentralized environmental governance are:

- access to social capital, including local knowledge, leaders and local shared vision;

- democratic access to information and decision-making;

- local government activity in environmental governance: as facilitator of access to natural resources, or as policy maker;

- an institutional framework that favours decentralized environmental governance and creates forums for social interaction and making widely accepted agreements acceptable.

The legitimacy of decisions depends on the local population's participation rate and on how well participants represent that population. With regard to public authorities, questions linked to biodiversity can be faced by adopting appropriate policies and strategies, through exchange of knowledge and experience, the forming of partnerships, correct management of land use, monitoring of biodiversity and optimal use of resources, or reducing consumption, and promoting environmental certifications, such as EMAS and/or ISO 14001. Local authorities undoubtedly have a central role to play in the protection of biodiversity and this strategy is successful above all when the authorities show strength by involving stakeholders in a credible environmental improvement project and activating a transparent and effective communication policy (Ioppolo et al., 2013).

State Governance

States play a crucial role in environmental governance, because "however far and fast international economic integration proceeds, political authority remains vested in national governments". It is for this reason that governments should respect and support the commitment to implementation of international agreements.

At the state level, environmental management has been found to be conducive to the creation of roundtables and committees. In France, the *Grenelle de l'environnement* process:

- included a variety of actors (e.g. the state, political leaders, unions, businesses, not-for-profit organizations and environmental protection foundations);

- allowed stakeholders to interact with the legislative and executive powers in office as indispensable advisors;

- worked to integrate other institutions, particularly the Economic and Social Council, to form a pressure group that participated in the process for creating an environmental governance model;

- attempted to link with environmental management at regional and local levels.

If environmental issues are excluded from e.g., the economic agenda, this may delegitimize those institutions.

"In southern countries, the main obstacle to the integration of intermediate levels in the process of territorial environmental governance development is often the dominance of developmentalist inertia in states' political mindset. The question of the environment has not been effectively integrated in national development planning and programs. Instead, the most common idea is that environmental protection curbs economic and social development, an idea encouraged by the frenzy for exporting raw materials extracted using destructive methods that consume resources and fail to generate any added value." Of course they are justified in this thinking, as their main concerns are social injustices such as poverty alleviation. Citizens in some of these states have responded by developing empowerment strategies to ease poverty through sustainable development. In addition to this, policymakers must be more aware of these concerns of the global south, and must make sure to integrate a strong focus on social justice in their policies.

Global Governance

At the global level there are numerous important actors involved in environmental governance and "a range of institutions contribute to and help define the practice of global environmental governance. The idea of global environmental governance is to govern the environment at a global level through a range of nation states and non state actors such as national governments, NGOs and other international organisations such as UNEP (United Nations Environment Programme). Global environmental governance is the answer to calls for new forms of governance because of the increasing complexity of the international agenda. It is perceived to be an effective form of multilateral management and essential to the international community in meeting goals of mitigation and the possible reversal of the impacts on the global environment. However, a precise definition of global environmental governance is still vague and there are many issues surrounding global governance. Elliot argues that "the congested institutional terrain still provides more of an appearance than a reality of comprehensive global governance." This meant that there are too many institutions within the global governance of the environment for it to be completely inclusive and coherent leaving it merely portraying the image of this to the global public. Global environmental governance is about more than simply expanding networks of institutions and decision makers. "It is a political practice which simultaneously reflects, constitutes and masks global relations of power and powerlessness." State agendas exploit the use of global environmental governance to enhance their oven agendas or wishes even if this is at the detriment of the vital element behind global environmental governance which is the environment. Elliot states that global environmental governance "is neither normatively neutral nor materially benign." As explored by Newell, report notes by The Global Environmental Outlook noted that the systems of global environmental governance are becoming increasingly irrelevant or impotent due to patterns of globalisation such as; imbalances in productivity and the distribution of goods and services, unsustainable progression of extremes of wealth and poverty and population and economic growth overtaking environmental gains. Newell states that, despite such acknowledgements, the "managing of global environmental change within International Relations continues to look to international regimes for the answers."

Issues of Scale

Multi-Tier Governance

The literature on governance scale shows how changes in the understanding of environmental issues have led to the movement from a local view to recognising their larger and more complicated scale. This move brought an increase in the diversity, specificity and complexity of initiatives. Meadowcroft pointed out innovations that were layered on top of existing structures and processes, instead of replacing them.

Lafferty and Meadowcroft give three examples of multi-tiered governance: internationalisation, increasingly comprehensive approaches, and involvement of multiple governmental entities. Lafferty and Meadowcroft described the resulting multi-tiered system as addressing issues on both smaller and wider scales.

Institutional Fit

Hans Bruyninckx claimed that a mismatch between the scale of the environmental problem and the level of the policy intervention was problematic. Young claimed that such mismatches reduced the effectiveness of interventions. Most of the literature addresses the level of governance rather than ecological scale.

Elinor Ostrom, amongst others, claimed that the mismatch is often the cause of unsustainable management practices and that simple solutions to the mismatch have not been identified.

Considerable debate has addressed the question of which level(s) should take responsibility for fresh water management. Development workers tend to address the problem at the local level. National governments focus on policy issues. This can create conflicts among states because rivers cross borders, leading to efforts to evolve governance of river basins.

Environmental Governance Issues

Soil Deterioration

Soil and land deterioration reduces its capacity for capturing, storing and recycling water, energy and food. Alliance 21 proposed solutions in the following domains:

- include soil rehabilitation as part of conventional and popular education

- involve all stakeholders, including policymakers and authorities, producers and land users, the scientific community and civil society to manage incentives and enforce regulations and laws

- establish a set of binding rules, such as an international convention

- set up mechanisms and incentives to facilitate transformations

- gather and share knowledge;

- mobilize funds nationally and internationally

Climate Change

The scientific consensus on climate change is expressed in the reports of Intergovernmental Panel on Climate Change (IPCC) and also in the statements by all major scientific bodies in the United States such as National Academy of Sciences.

The drivers of climate change can include - Changes in solar irradiance - Changes in atmospheric trace gas and aerosol concentrations Evidence of climate change can be identified by examining - Atmospheric concentrations of Green House Gases (GHGs) such as carbon dioxide (CO2) - Land and sea surface temperatures - Atmospheric water vapor - Precipitation - The occurrence or strength of extreme weather and climate events - Glaciers - Rapid sea ice loss - Sea level

It is suggested by climate models that the changes in temperature and sea level can be the causal effects of human activities such as consumption of fossil fuels, deforestation, increased agricultural production and production of xenobiotic gases.

There has been increasing actions in order to mitigate climate change and reduce its impact at national, regional and international levels. Kyoto protocol and United Nations Framework Convention on Climate Change (UNFCCC) plays the most important role in addressing climate change at an international level.

The goal of combating climate change led to the adoption of the Kyoto Protocol by 191 states, an agreement encouraging the reduction of greenhouse gases, mainly CO_2. Since developed economies produce more emissions per capita, limiting emissions in all countries inhibits opportunities for emerging economies, the only major success in efforts to produce a global response to the phenomenon.

Two decades following the Brundtland Report, however, there has been no improvement in the key indicators highlighted.

Biodiversity

Environmental governance for protecting the biodiversity has to act in many levels. Biodiversity is fragile because it is threatened by almost all human actions. To promote conservation of biodiversity, agreements and laws have to be created to regulate agricultural activities, urban growth, industrialization of countries, use of natural resources, control of invasive species, the correct use of water and protection of air quality. Before making any decision for a region or country decision makers, politicians and community have to take into account what are the potential impacts for biodiversity, that any project can have.

Population growth and urbanization have been a great contributor for deforestation. Also, population growth requires more intense agricultural areas use, which also results in necessity of new areas to be deforested. This causes habitat loss, which is one of the major threats for biodiversity. Habitat loss and habitat fragmentation affects all species, because they all rely on limited resources, to feed on and to breed.

'Species are genetically unique and irreplaceable their loss is irreversible. Ecosystems vary across a vast range of parameters, and similar ecosystems (whether wetlands, forests, coastal *reserves etc) cannot be presumed to be interchangeable, such that the loss of one can be compensated by protection or restoration of another'.*

To avoid habitat loss, and consequently biodiversity loss, politicians and lawmakers should be aware of the precautionary principle, which means that before approving a project or law all the pros and cons should be carefully analysed. Sometimes the impacts are not explicit, or not even proved to exist. However, if there is any chance of an irreversible impact happen, it should be taken into consideration.

To promote environmental governance for biodiversity protection there has to be a clear articulation between values and interests while negotiating environmental management plans. International agreements are good way to have it done right.

The Convention on Biological Diversity (CBD) was signed in Rio de Janeiro in 1992 human activities. The CBD's objectives are: "to conserve biological diversity, to use biological diversity in a sustainable fashion, to share the benefits of biological diversity fairly and equitably." The Convention is the first global agreement to address all aspects of biodiversity: genetic resources, species and ecosystems. It recognizes, for the first time, that the conservation of biological diversity is "a common concern for all humanity". The Convention encourages joint efforts on measures for scientific and technological cooperation, access to genetic resources and the transfer of clean environmental technologies.

The Convention on Biological Diversity most important edition happened in 2010 when the Strategic Plan for Biodiversity 2011-2020 and the Aichi Targets, were launched. These two projects together make the United Nations decade on Biodiversity. It was held in Japan and has the targets of *'halting and eventually reversing the loss of biodiversity of the planet'*. The Strategic Plan for Biodiversity has the goal to *'promote its overall vision of living in harmony with nature'* As result *'mainstream biodiversity at different levels. Throughout the United Nations Decade on Biodiversity, governments are encouraged to develop, implement and communicate the results of national strategies for implementation of the Strategic Plan for Biodiversity'*. According to the CBD the five Aichi targets are:

- *'Address the underlying causes of biodiversity loss by mainstreaming biodiversity across government and society*;

- *Reduce the direct pressures on biodiversity and promote sustainable use*;

- *Improve the status of biodiversity by safeguarding ecosystems, species and genetic diversity*;

- *Enhance the benefits to all from biodiversity and ecosystem services*;

- *Enhance implementation through participatory planning, knowledge management and capacity building.'*

Water

The 2003 UN World Water Development Report claimed that the amount of water available over the next twenty years would drop by 30%.

At that time, 40% of the planet's inhabitants did not have access to the minimum necessary for basic hygiene. Over 2.2 million people died in 2000 from diseases linked to contaminated water, or from drowning. In 2004, the UK's WaterAid charity reported that one child died every 15 seconds from water-linked diseases.

According to Alliance 21 "All levels of water supply management are necessary and independent. The integrated approach to the catchment areas must take into account the needs of irrigation and those of towns, jointly and not separately as is often seen to be the case....The governance of a water supply must be guided by the principles of sustainable development."

Australian water resources have always been variable but they are becoming increasingly so with changing climate conditions. Because of how limited water resources are in Australia, there needs to be an effective implementation of environmental governance conducted within the country. Water restrictions are an important policy device used in Australian environmental governance to limit the amount of water used in urban and agricultural environments (Beeton et al. 2006). There is increased pressure on surface water resources in Australia because of the uncontrolled growth in groundwater use and the constant threat of drought. These increased pressures not only affect the quantity and quality of the waterways but they also negatively affect biodiversity. The government needs to create policies that preserve, protect and monitor Australia's inland water. The most significant environmental governance policy imposed by the Australian government is environmental flow allocations that allocate water to the natural environment. The proper implementation of water trading systems could help to conserve water resources in Australia. Over the years there has been an increase in demand for water, making Australia the third largest per capita user of water in the world (Beeton et al. 2006). If this trend continues, the gap between supply and demand will need to be addressed. The government needs to implement more efficient water allocations and raise water rates (UNEP, 2014). By changing public perception to promote the action of reusing and recycling water some of the stress of water shortages can be alleviated. More extensive solutions like desalination plants, building more dams and using aquifer storage are all options that could be taken to conserve water levels but all these methods are controversial. With caps on surface water use, both urban and rural consumers are turning to groundwater use; this has caused groundwater levels to decline significantly. Groundwater use is very hard to monitor and regulate. There is not enough research currently being conducted to accurately determine sustainable yields. Some regions are seeing improvement in groundwater levels by applying caps on bores and the amount of water that consumers are allowed to extract. There have been projects in environmental governance aimed at restoring vegetation in the riparian zone. Restoring riparian vegetation helps increase biodiversity, reduce salinity, prevent soil erosion and prevent riverbank collapse. Many rivers and waterways are controlled by weirs and locks that control the flow of rivers and also prevent the movement of fish. The government has funded fish-ways on some weirs and locks to allow for native fish to move upstream. Wetlands have significantly suffered under restricted water resources with water bird numbers dropping and a decrease in species diversity. The allocation of water for bird breeding through environmental flows in Macquarie Marshes has led to an increase in breeding (Beeton et al. 2006). Because of dry land salinity throughout Australia there has been an increase in the levels of salt in Australian waterways. There has been funding in salt interception schemes which help to improve in-stream salinity levels but whether river salinity has improved or not is still unclear because there is not enough data available yet. High salinity levels are dangerous because they can negatively affect larval and juvenile stages of certain fish.

The introduction of invasive species into waterways has negatively affected native aquatic species because invasive species compete with native species and alter natural habitats. There has been research in producing daughterless carp to help eradicate carp. Government funding has also gone into building in-stream barriers that trap the carp and prevent them from moving into floodplains and wetlands. Investment in national and regional programmes like the Living Murray (MDBC), Healthy Waterways Partnership and the Clean Up the Swan Programme are leading to important environmental governance. The Healthy Rivers programme promotes restoration and recovery of environmental flows, riparian re-vegetation and aquatic pest control. The Living Murray programme has been crucial for the allocation of water to the environment by creating an agreement to recover 500 billion litres of water to the Murray River environment. Environmental governance and water resource management in Australia must be constantly monitored and adapted to suit the changing environmental conditions within the country (Beeton et al. 2006). If environmental programmes are governed with transparency there can be a reduction in policy fragmentation and an increase in policy efficiency (McIntyre, 2010).

Ozone Layer

On 16 September 1987 the United Nations General Assembly signed the Montreal Protocol to address the declining ozone layer. Since that time, the use of chlorofluorocarbons (industrial refrigerants and aerosols) and farming fungicides such as methyl bromide has mostly been eliminated, although other damaging gases are still in use.

Nuclear Risk

The Nuclear non-proliferation treaty is the primary multilateral agreement governing nuclear activity.

Transgenic Organisms

Genetically modified organisms are not the subject of any major multilateral agreements. They are the subject of various restrictions at other levels of governance. GMOs are in widespread use in the US, but are heavily restricted in many other jurisdictions.

Controversies have ensued over golden rice, genetically modified salmon, genetically modified seeds, disclosure and other topics.

Precautionary Principle

The precautionary principle or precautionary approach states that if an action or policy has a suspected risk of causing harm to the public or to the environment, in the absence of scientific consensus that the action or policy is harmful, the burden of proof that it is not harmful falls on those taking an action. As of 2013 it was not the basis of major multilateral agreements. The Precautionary Principle is put into effect if there is a chance that proposed action may cause harm to the society or the environment. Therefore, those involved in the proposed action must provide evidence that it will not be harmful, even if scientists do not believe that it will cause harm. It falls upon the policymakers to make the optimal decision, if there is any risk, even without any credible scientific evidence. However, taking precautionary action also means that there is an element of cost involved, either social or economic. So if the cost was seen as insignificant the action would be taken without the implementation of the precautionary principle. But often the cost is ignored, which can lead to harmful repercussions. This is often the case with industry and scientists who are primarily concerned with protecting their own interests.

Agreements

Conventions

The main multilateral conventions, also known as Rio Conventions, are as follows:

Convention on Biological Diversity (CBD) (1992–1993): aims to conserve biodiversity. Related agreements include the Cartagena Protocol on biosafety.

United Nations Framework Convention on Climate Change (UNFCC) (1992–1994): aims to stabilize concentrations of greenhouse gases at a level that would stabilize the climate system without threatening food production, and enabling the pursuit of sustainable economic development; it incorporates the Kyoto Protocol.

United Nations Convention to Combat Desertification (UNCCD) (1994–1996): aims to combat desertification and mitigate the effects of drought and desertification, particularly in Africa.

Further Conventions:

- Ramsar Convention on Wetlands of International Importance (1971–1975)
- UNESCO World Heritage Convention (1972–1975)
- Convention on International Trade in Endangered Species of Wild Flora and Fauna (CITES) (1973–1975)
- Bonn Convention on the Conservation of Migratory Species (1979–1983)
- Convention on the Protection and Use of Transboundary Watercourses and International Lakes (Water Convention) (1992–1996)
- Basel Convention on the Control of Transboundary Movements of Hazardous Wastes and their Disposal (1989–1992)
- Rotterdam Convention on the Prior Informed Consent Procedures for Certain Hazardous Chemicals and Pesticides in International Trade
- Stockholm Convention on Persistent Organic Pollutants (COP) (2001–2004)

The Rio Conventions are Characterized by:

- obligatory execution by signatory states
- involvement in a sector of global environmental governance

- focus on the fighting poverty and the development of sustainable living conditions;

- funding from the Global Environment Facility (GEF) for countries with few financial resources;

- inclusion of a for assessing ecosystem status

Environmental Conventions are Regularly Criticized for Their:

- rigidity and verticality: they are too descriptive, homogenous and top down, not reflecting the diversity and complexity of environmental issues. Signatory countries struggle to translate objectives into concrete form and incorporate them consistently;

- duplicate structures and aid: the sector-specific format of the conventions produced duplicate structures and procedures. Inadequate cooperation between government ministries;

- contradictions and incompatibility: e.g., "if reforestation projects to reduce CO_2 give preference to monocultures of exotic species, this can have a negative impact on biodiversity (whereas natural regeneration can strengthen both biodiversity and the conditions needed for life)."

Until now, the formulation of environmental policies at the international level has been divided by theme, sector or territory, resulting in treaties that overlap or clash. International attempts to coordinate environment institutions, include the Inter-Agency Coordination Committee and the Commission for Sustainable Development, but these institutions are not powerful enough to effectively incorporate the three aspects of sustainable development.

Multilateral Environmental Agreements (MEAs)

MEAs are agreements between several countries that apply internationally or regionally and concern a variety of environmental questions. As of 2013 over 500 Multilateral Environmental Agreements (MEAs), including 45 of global scope involve at least 72 signatory countries. Further agreements cover regional environmental problems, such as deforestation in Borneo or pollution in the Mediterranean. Each agreement has a specific mission and objectives ratified by multiple states.

Many Multilateral Environmental Agreements have been negotiated with the support from the United Nations Environmental Programme and work towards the achievement of the United Nations Millennium Development Goals as a means to instil sustainable practices for the environment and its people. Multilateral Environmental Agreements are considered to present enormous opportunities for greener societies and economies which can deliver numerous benefits in addressing food, energy and water security and in achieving sustainable development. These agreements can be implemented on a global or regional scale, for example the issues surrounding the disposal of hazardous waste can be implemented on a regional level as per the Bamako Convention on the Ban of the Import into Africa and the Control of Transboundary Movement and Management of Hazardous Waste within Africa which applies specifically to Africa, or the global approach to hazardous waste such as the Basel Convention on the Control of Transboundary Movements of Hazardous Wastes and their Disposal which is monitored throughout the world.

"The environmental governance structure defined by the Rio and Johannesburg Summits is sustained by UNEP, MEAs and developmental organizations and consists of assessment and policy development, as well as project implementation at the country level.

"The Governance Structure Consists of a Chain of Phases:

- a) assessment of environment status;

- b) international policy development;

- c) formulation of MEAs;

- d) policy implementation;

- e) policy assessment;

- f) enforcement;

- g) sustainable development.

"Traditionally, UNEP has focused on the normative role of engagement in the first three phases. Phases (d) to (f) are covered by MEAs and the sustainable development phase involves developmental organizations such as UNDP and the World Bank."

Lack of coordination affects the development of coherent governance. The report shows that donor states support development organizations, according to their individual interests. They do not follow a joint plan, resulting in overlaps and duplication. MEAs tend not to become a joint frame of reference and therefore receive little financial support. States and organizations emphasize existing regulations rather than improving and adapting them.

Background

The risks associated with nuclear fission raised global awareness of environmental threats. The 1963 Partial Nuclear Test Ban Treaty prohibiting atmospheric nuclear testing was the beginning of the globalization of environmental issues. Environmental law began to be modernized and co-ordinated with the Stockholm Conference (1972), backed up in 1980 by the Vienna Convention on the Law of Treaties. The Vienna Convention for the Protection of the Ozone Layer was signed and ratified in 1985. In 1987, 24 countries signed the Montreal Protocol which imposed the gradual withdrawal of CFCs.

The Brundtland Report, published in 1987 by the UN Commission on Environment and Development, stipulated the need for economic development that "meets the needs of the present without compromising the capacity of future generations to meet their needs.

Rio Conference (1992) and Reactions

The United Nations Conference on Environment and Development (UNCED), better known as the 1992 Earth Summit, was the first major international meeting since the end of the Cold War and was attended by delegations from 175 countries. Since then the biggest international conferences that take place every 10 years guided the global governance process with a series of MEAs. Environmental treaties are applied with the help of secretariats.

Governments created international treaties in the 1990s to check global threats to the environment. These treaties are far more restrictive than global protocols and set out to change non-sustainable production and consumption models.

Agenda 21

Agenda 21 is a detailed plan of actions to be implemented at the global, national and local levels by UN organizations, member states and key individual groups in all regions. Agenda 21 advocates making sustainable development a legal principle law. At the local level, local Agenda 21 advocates an inclusive, territory-based strategic plan, incorporating sustainable environmental and social policies.

The Agenda has been accused of using neoliberal principles, including free trade to achieve environmental goals. For example, chapter two, entitled "International Cooperation to Accelerate Sustainable Development in Developing Countries and Related Domestic Policies" states, "The international economy should provide a supportive international climate for achieving environment and development goals by: promoting sustainable development through trade liberalization."

Actors

International Institutions

United Nations Environment Program

The UNEP has had its biggest impact as a monitoring and advisory body, and in developing environmental agreements. It has also contributed to strengthening the institutional capacity of environment ministries.

In 2002 UNEP held a conference to focus on product lifecycle impacts, emphasizing the fashion, advertising, financial and retail industries, seen as key agents in promoting sustainable consumption.

According to Ivanova, UNEP adds value in environmental monitoring, scientific assessment and information sharing, but cannot lead all environmental management processes. She proposed the following tasks for UNEP:

- initiate a strategic independent overhaul of its mission;
- consolidate the financial information and transparency process;
- restructure organizing governance by creating an operative executive council that balances the omnipresence of the overly imposing and fairly ineffectual Governing Council/Global Ministerial Environment Forum (GMEF).

Other proposals offer a new mandate to "produce greater unity amongst social and environmental agencies, so that the concept of 'environment for development' becomes a reality. It needs to act as a platform for establishing standards and for other types of interaction with national and international organizations and the United Nations. The principles of cooperation and common but differentiated responsibilities should be reflected in the application of this revised mandate."

Sherman proposed principles to strengthen UNEP:

- obtain a social consensus on a long-term vision;

- analyze the current situation and future scenarios;

- produce a comprehensive plan covering all aspects of sustainable development;

- build on existing strategies and processes;

- multiply links between national and local strategies;

- include all these points in the financial and budget plan;

- adopt fast controls to improve process piloting and identification of progress made;

- implement effective participation mechanisms.

Another group stated, "Consider the specific needs of developing countries and respect of the fundamental principle of 'common but differentiated responsibilities'. Developed countries should promote technology transfer, new and additional financial resources, and capacity building for meaningful participation of developing countries in international environmental governance. Strengthening of international environmental governance should occur in the context of sustainable development and should involve civil society as an important stakeholder and agent of transformation."

Global Environment Facility (GEF)

Created in 1991, the Global Environment Facility is an independent financial organization initiated by donor governments including Germany and France. It was the first financial organization dedicated to the environment at the global level. As of 2013 it had 179 members. Donations are used for projects covering biodiversity, climate change, international waters, destruction of the ozone layer, soil degradation and persistent organic pollutants.

GEF's institutional structure includes UNEP, UNDP and the World Bank. It is the funding mechanism for the four environmental conventions: climate change, biodiversity, persistent organic pollutants and desertification. GEF transfers resources from developed countries to developing countries to fund UNDP, UNEP and World Bank projects. The World Bank manages the annual budget of US$561.10 million.

The GEF has been criticized for its historic links with the World Bank, at least during its first phase during the 1990s, and for having favoured certain regions to the detriment of others. Another view sees it as contributing to the emergence of a global "green market". It represents "an adaptation (of the World Bank) to this emerging world order, as a response to the emergence of environmental movements that are becoming a geopolitical force." Developing countries demanded financial transfers to help them protect their environment.

GEF is subject to economic profitability criteria, as is the case for all the conventions. It received more funds in its first three years than the UNEP has since its creation in 1972. GEF funding represents less than 1% of development aid between 1992 and 2002.

United Nations Commission on Sustainable Development (CSD)

This intergovernmental institution meets twice a year to assess follow-up on Rio Summit goals. The CSD is made up of 53 member states, elected every three years and was reformed in 2004 to help improve implementation of Agenda 21. It meets twice a year, focusing on a specific theme during each two-year period: 2004-2005 was dedicated to water and 2006-2007 to climate change. The CSD has been criticized for its low impact, general lack of presence and the absence of Agenda 21 at the state level specifically, according to a report by the World Resources Institute. Its mission focuses on sequencing actions and establishing agreements puts it in conflict with institutions such as UNEP and OECD.

World Environment Organization (WEO)

A proposed World Environment Organization, analogous to the World Health Organization could be capable of adapting treaties and enforcing international standards.

The European Union, particularly France and Germany, and a number of NGOs favour creating a WEO. The United Kingdom, the US and most developing countries prefer to focus on voluntary initiatives. WEO partisans maintain that it could offer better political leadership, improved legitimacy and more efficient coordination. Its detractors argue that existing institutions and missions already provide appropriate environmental governance; however the lack of coherence and coordination between them and the absence of clear division of responsibilities prevents them from greater effectiveness.

World Bank

The World Bank influences environmental governance through other actors, particularly the GEF. The World Bank's mandate is not sufficiently defined in terms of environmental governance despite the fact that it is included in its mission. However, it allocates 5 to 10% of its annual funds to environmental projects. The institution's capitalist vocation means that its investment is concentrated solely in areas which are profitable in terms of cost benefits, such as climate change action and ozone layer protection, whilst neglecting other such as adapting to climate change and desertification. Its financial autonomy means that it can make its influence felt indirectly on the creation of standards, and on international and regional negotiations.

Following intense criticism in the 1980s for its support for destructive projects which, amongst other consequences, caused deforestation of tropical forests, the World Bank drew up its own environment-related standards in the 1990s so it could correct its actions. These standards differ from UNEP's standards, meant to be the benchmark, thus discrediting the institution and sowing disorder and conflict in the world of environmental governance. Other financial institutions, regional development banks and the private sector also drew up their own standards. Criticism is not directed at the World Bank's standards in themselves, which Najam considered as "robust", but at their legitimacy and efficacy.

GEF

The GEF's account of itself as of 2012 is as "the largest public funder of projects to improve the global environment", period, which "provides grants for projects related to biodiversity, climate change, international waters, land degradation, the ozone layer, and persistent organic pollutants." It claims to have provided "$10.5 billion in grants and leveraging $51 billion in co-financing for over 2,700 projects in over 165 countries [and] made more than 14,000 small grants directly to civil society and community-based organizations, totaling $634 million." It serves as mechanism for the:

- Convention on Biological Diversity (CBD)

- United Nations Framework Convention on Climate Change (UNFCCC)

- Stockholm Convention on Persistent Organic Pollutants (POPs)

- Convention to Combat Desertification (UNCCD)

- implementation of Montreal Protocol on Substances That Deplete the Ozone Layer in some countries with "economies in transition"

This mandate reflects the restructured GEF as of October 2011 .

World Trade Organization (WTO)

The WTO's mandate does not include a specific principle on the environment. All the problems linked to the environment are treated in such a way as to give priority to trade requirements and the principles of the WTO's own trade system. This produces conflictual situations. Even if the WTO recognizes the existence of MEAs, it denounces the fact that around 20 MEAs are in conflict with the WTO's trade regulations. Furthermore, certain MEAs can allow a country to ban or limit trade in certain products if they do not satisfy established environmental protection requirements. In these circumstances, if one country's ban relating to another country concerns two signatories of the same MEA, the principles of the treaty can be used to resolve the disagreement, whereas if the country affected by the trade ban with another country has not signed the agreement, the WTO demands that the dispute be resolved using the WTO's trade principles, in other words, without taking into account the environmental consequences.

Some criticisms of the WTO mechanisms may be too broad. In a recently dispute over labelling of dolphin safe labels for tuna between the US and Mexico, the ruling was relatively narrow and did not, as some critics claimed,

International Monetary Fund (IMF)

The IMF's mission is to encourage growth and development. The IMF advocates reduced public expenditure, increased exports and foreign investment. The environment is not a priority for the IMF, leading it to favor projects that may have negative environmental effects.

The IMF Green Fund proposal of Dominique Strauss-Kahn specifically to address "climate-related shocks in Africa", despite receiving serious attention was rejected. Strauss-Kahn's proposal, backed by France and Britain, was that "developed countries would make an initial capital injection into the fund using some of the $176 billion worth of SDR allocations from last year in exchange for a stake in the green fund." However, "most of the 24 directors ... told Strauss-Kahn that climate was not part of the IMF's mandate and that SDR allocations are a reserve asset never intended for development issues."

UN ICLEI

The UN's main body for coordinating municipal and urban decision-making is named the International Council for Local Environmental Initiatives. Its slogan is "Local Governments for Sustainability". This body sponsored the concept of full cost accounting that makes environmental governance the foundation of other governance.

ICLEIs projects and achievements include:

- Convincing thousands of municipal leaders to sign the World Mayors and Municipal Leaders Declaration on Climate Change (2005) which notably requests of other levels of government that:

- Global trade regimes, credits and banking reserve rules be reformed to advance debt relief and incentives to implement polices and practices that reduce and mitigate climate change.

- Starting national councils to implement this and other key agreements, e.g., ICLEI Local Governments for Sustainability USA

- Spreading ecoBudget (2008) and Triple Bottom Line (2007) "tools for embedding sustainability into council operations", e.g. Guntur's Municipal Corporation, one of the first four to ipmlement the entire framework.

- Sustainability Planning Toolkit (launched 2009) integrating these and other tools

- Cities Climate Registry (launched 2010) - part of UNEP Campaign on Cities and Climate Change

ICLEI promotes best practice exchange among municipal governments globally, especially green infrastructure, sustainable procurement.

Other Secretariats

Other international institutions incorporate environmental governance in their action plans, including:

- United Nations Development Programme (UNDP), promoting development;

- World Meteorological Organization (WMO) which works on the climate and atmosphere;

- Food and Agriculture Organisation (FAO) working on the protection of agriculture, forests and fishing;

- International Atomic Energy Agency (IAEA) which focuses on nuclear security.

Over 30 UN agencies and programmes support environmental management, according to Najam. This produces a lack of coordination, insufficient exchange of information and dispersion of responsibilities. It also results in proliferation of initiatives and rivalry between them.

Criticism

According to Bauer, Busch and Siebenhüner, the different conventions and multilateral agreements of global environmental regulation is increasing their secretariats' influence. Influence varies according to bureaucratic and leadership efficiency, choice of technical or client-centered.

The United Nations is often the target of criticism, including from within over the multiplication of secretariats due to the chaos it produces. Using a separate secretariat for each MEA creates enormous overhead given the 45 international-scale and over 500 other agreements.

States

Environmental Governance at the State Level

Environmental protection has created opportunities for mutual and collective monitoring among neighbouring states. The European Union provides an example of the institutionalization of joint regional and state environmental governance. Key areas include information, led by the European Environment Agency (EEA), and the production and monitoring of norms by states or local institutions.

State Participation in Global Environmental Governance

US refusal to ratify major environment agreements produced tensions with ratifiers in Europe and Japan.

The World Bank, IMF and other institutions are dominated by the developed countries and do not always properly consider the requirements of developing countries.

Business

Environmental governance applies to business as well as government. Considerations are typical of those in other domains:

- values (vision, mission, principles);

- policy (strategy, objectives, targets);

- oversight (responsibility, direction, training, communication);

- process (management systems, initiatives, internal control, monitoring and review, stakeholder dialogue, transparency, environmental accounting, reporting and verification);

- performance (performance indicators, benchmarking, eco-efficiency, reputation, compliance, liabilities, business development).

White and Klernan among others discuss the correlation between environmental governance and financial performance. This correlation is higher in sectors where environmental impacts are greater.

Business environmental issues include emissions, biodiversity, historical liabilities, product and material waste/recycling, energy use/supply and many others.

Environmental governance has become linked to traditional corporate governance as an increasing number of shareholders are corporate environmental impacts. Corporate governance is the set of processes, customs, policies, laws, and institutions affecting the way a corporation (or company) is managed. Corporate governance is affected by the relationships among stakeholders. These stakeholders research and quantify performance to compare and contrast the environmental performance of thousands of companies.

Large corporations with global supply chains evaluate the environmental performance of business partners and suppliers for marketing and ethical reasons. Some consumers seek environmentally friendly and sustainable products and companies.

Non-Governmental Organizations

According to Bäckstrand and Saward, "broader participation by non-state actors in multilateral environmental decisions (in varied roles such as agenda setting, campaigning, lobbying, consultation, monitoring, and implementation) enhances the democratic legitimacy of environmental governance."

Local activism is capable of gaining the support of the people and authorities to combat environmental degradatation. In Cotacachi, Ecuador, a social movement used a combination of education, direct action, the influence of local public authorities and denunciation of the mining company's plans in its own country, Canada, and the support of international environmental groups to influence mining activity.

Fisher cites cases in which multiple strategies were used to effect change. She describes civil society groups that pressure international institutions and also organize local events. Local groups can take responsibility for environmental governance in place of governments.

According to Bengoa, "social movements have contributed decisively to the creation of an institutional platform wherein the fight against poverty and exclusion has become an inescapable benchmark." But despite successes in this area, "these institutional changes have not produced the processes for transformation that could have made substantial changes to the opportunities available to rural inhabitants, particularly the poorest and those excluded from society." He cites several reasons:

- conflict between in-group cohesion and openness to outside influence;
- limited trust between individuals;
- contradiction between social participation and innovation;
- criticisms without credible alternatives to environmentally damaging activities

A successful initiative in Ecuador involved the establishment of stakeholder federations and management committees (NGOs, communities, municipalities and the ministry) for the management of a protected forest.

Proposals

The International Institute for Sustainable Development proposed an agenda for global governance. These objectives are:

- expert leadership;
- positioning science as the authoritative basis of sound environmental policy;
- coherence and reasonable coordination;
- well-managed institutions;
- incorporate environmental concerns and actions within other areas of international policy and action

Coherence and Coordination

Despite the increase in efforts, actors, agreements and treaties, the global environment continue to degrade at a rapid rate. From the big hole in Earth's ozone layer to over-fishing to the uncertainties of climate change, the world is confronted by several intrinsically global challenges. However, as the environmental agenda becomes more complicated and extensive, the current system has proven ineffective in addressing and tackling problems related to trans-boundary externalities and the environment is still experiencing degradation at unprecedented levels.

Inforesources identifies four major obstacles to global environmental governance, and describes measures in response. The four obstacles are:

- parallel structures and competition, without a coherent strategy
- contradictions and incompatibilities, without appropriate compromise
- competition between multiple agreements with incompatible objectives, regulations and processes
- integrating policy from macro- to micro- scales.

Recommended Measures:

- MDGs (Milenium Development Goals) and conventions, combining sustainability and reduction of poverty and equity;
- country-level approach linking global and local scales
- coordination and division of tasks in a multilateral approach that supports developing countries and improves coordination between donor countries and institutions
- use of Poverty Reduction Strategy Papers (PRSPs) in development planning
- transform conflicts into tradeoffs, synergies and win-win options

Contemporary debates surrounding global environmental governance have converged on the idea of developing a stronger and more effective institutional framework. The views on how to achieve this, however, still hotly debated. Currently, rather than teaming up with the United Nations Environment Programme (UNEP), international environmental responsibilities have been spread across many different agencies including: a) specialised agencies within the UN system such as the World Meteorological Organisation, the International Maritime Organisation and others; b) the programs in the UN system such as the UN Development Program; c) the UN regional economic and social commission; d) the Bretton Woods institutions; e) the World Trade Organisation and; f) the environmentally focused mechanisms such as the Global Environment Facility and close to 500 international environmental agreements.

Some analysts also argue that multiple institutions and some degree of overlap and duplication in policies is necessary to ensure maximum output from the system. Others, however, claim that institutions have become too dispersed and lacking in coordination which can be damaging to their effectiveness in global environmental governance. Whilst there are various arguments for and against a WEO, the key challenge, however, remains the same: how to develop a rational and effective framework that will protect the global environment efficiently.

Democratization

Starting in 2002, Saward and others began to view the Earth Summit process as capable opening up the possibility of stakeholder democracy. The summits were deliberative rather than simply participative, with NGOs, women, men, indigenous peoples and businesses joining the decision-making process alongside states and international organizations, characterized by:

- the importance given to scientific and technical considerations

- the official and unofficial participation of many actors with heterogeneous activity scopes

- growing uncertainty

- a new interpretation of international law and social organization models

As of 2013, the absence of joint rules for composing such fora leads to the development of non-transparent relations that favour the more powerful stakeholders. Criticisms assert that they act more as a lobbying platform, wherein specific interest groups attempt to influence governments.

Institutional Reform

Actors inside and outside the United Nations are discussing possibilities for global environmental governance that provides a solution to current problems of fragility, coordination and coherence. Deliberation is focusing on the goal of making UNEP more efficient. A 2005 resolution recognizes "the need for more efficient environmental activities in the United Nations system, with enhanced coordination, improved policy advice and guidance, strengthened scientific knowledge, assessment and cooperation, better treaty compliance, while respecting the legal autonomy of the treaties, and better integration of environmental activities in the broader sustainable development framework."

Proposals Include:

- greater and better coordination between agencies;

- strengthen and acknowledge UNEP's scientific role;

- identify MEA areas to strengthen coordination, cooperation and teamwork between different agreements;

- increase regional presence;

- implement the Bali Strategic Plan on improving technology training and support for the application of environmental measures in poor countries;

- demand that UNEP and MEAs participate formally in all relevant WTO committees as observers.

- strengthen its financial situation;

- improve secretariats' efficiency and effectiveness.

One of the Main Studies Addressing this Issue Proposes:

- clearly divide tasks between development organizations, UNEP and the MEAs

- adopt a political direction for environmental protection and sustainable development

- authorize the UNEP Governing Council/Global Ministerial Environment Forum to adopt the UNEP medium-term strategy

- allow Member States to formulate and administer MEAs an independent secretariat for each convention

- support UNEP in periodically assessing MEAs and ensure coordination and coherence

- establish directives for setting up national/regional platforms capable of incorporating MEAs in the Common Country Assessment (CCA) process and United Nations Development Assistance Framework (UNDAF)

- establish a global joint planning framework

- study the aptitude and efficiency of environmental activities' funding, focusing on differential costs

- examine and redefine the concept of funding differential costs as applicable to existing financial mechanisms

- reconsider remits, division of tasks and responsibilities between entities that provide services to the multipartite conferences. Clearly define the services that UN offices provide to MEA secretariats

- propose measures aiming to improve personnel provision and geographic distribution for

MEA secretariats

- improve transparency resource use for supporting programmes and in providing services to MEAs. Draw up a joint budget for services supplied to MEAs.

Education

A 2001 Alliance 21 report proposes six fields of action:

- strengthen citizens' critical faculties to ensure greater democratic control of political orientations

- develop a global and critical approach

- develop civic education training for teachers

- develop training for certain socio-professional groups

- develop environmental education for the entire population;

- assess the resulting experiences of civil society

Transform Daily Life

Individuals can modify consumption, based on voluntary simplicity: changes in purchasing habits, simplified lifestyles (less work, less consumption, more socialization and constructive leisure time). But individual actions must not replace vigilance and pressure on policies. Notions of responsible consumption developed over decades, revealing the political nature of individual purchases, according to the principle that consumption should satisfy the population's basic needs. These needs comprise the physical wellbeing of individuals and society, a healthy diet, access to drinking water and plumbing, education, healthcare and physical safety. The general attitude centres on the need to reduce consumption and reuse and recycle materials. In the case of food consumption, local, organic and fair trade products which avoid ill treatment of animals has become a major trend.

Alternatives to the personal automobile are increasing, including public transport, car sharing and bicycles and alternative propulsion systems.

Alternative energy sources are becoming less costly.

Ecological industrial processes turn the waste from one industry into raw materials for another.

Governments can reduce subsidies/increase taxes/tighten regulation on unsustainable activities.

The Community Environmental governance Global Alliance encourages holistic approaches to environmental and economic challenges, incorporating indigenous knowledge. Okotoks, Alberta capped population growth based on the carrying capacity of the Sheep River. The Fraser Basin Council Watershed Governance in British Columbia, Canada, manages issues that span municipal jurisdictions. Smart Growth is an international movement that employs key tenets of Environmental governance in urban planning.

Policies and Regulations

Establish policies and regulations that promote "infrastructures for well being" whilst addressing the political, physical and cultural levels.

Eliminate subsidies that have a negative environmental impact and tax pollution

Promoting workers' personal and family development.

Coordination

A programme of national workshops on synergies between the three Rio Conventions launched in late 2000, in collaboration with the relevant secretariats. The goal was to strengthen coordination at the local level by:

- sharing information

- promoting political dialogue to obtain financial support and implement programmes

- enabling the secretariats to update their joint work programmes.

According to Campbell, "In the context of globalization, the question of linking up environmental themes with other subjects, such as trade, investment and conflict resolution mechanisms, as well as the economic incentives to participate in and apply agreements would seem to provide an important lesson for the effective development of environmental governance structures." Environmental concerns would become part of the global economic system. "These problems also contain the seeds of a new generation of international conflicts that could affect both the stability of international relations and collective security. Which is why the concept of 'collective security' has arisen."

Moving local decisions to the global level is as important as the way in which local initiatives and best practices are part of a global system. Kanie points out that NGOs, scientists, international institutions and stakeholder partnerships can reduce the distance that separates the local and international levels.

Environmental Movement

The environmental movement (sometimes referred to as the ecology movement), also including conservation and green politics, is a diverse scientific, social, and political movement for addressing environmental issues. Environmentalists advocate the sustainable management of resources and stewardship of the environment through changes in public policy and individual behavior. In its recognition of humanity as a participant in (not enemy of) ecosystems, the movement is centered on ecology, health, and human rights.

The environmental movement is an international movement, represented by a range of organizations, from the large to grassroots and varies from country to country. Due to its large membership, varying and strong beliefs, and occasionally speculative nature, the environmental movement is not always united in its goals.

Apollo 8's *Earthrise*, 24 December 1968

The movement also encompasses some other movements with a more specific focus, such as the climate movement. At its broadest, the movement includes private citizens, professionals, religious devotees, politicians, scientists, nonprofit organizations and individual advocates.

History

Early Awareness

Levels of air pollution rose during the Industrial Revolution, sparking the first modern environmental laws to be passed in the mid-19th century.

Early interest in the environment was a feature of the Romantic movement in the early 19th century. The poet William Wordsworth had travelled extensively in the Lake District and wrote that it is a "sort of national property in which every man has a right and interest who has an eye to perceive and a heart to enjoy".

The origins of the environmental movement lay in the response to increasing levels of smoke pollution in the atmosphere during the Industrial Revolution. The emergence of great factories and the concomitant immense growth in coal consumption gave rise to an unprecedented level of air pollution in industrial centers; after 1900 the large volume of industrial chemical discharges added to the growing load of untreated human waste. Under increasing political pressure from the urban middle-class, the first large-scale, modern environmental laws came in the form of Britain's Alkali Acts, passed in 1863, to regulate the deleterious air pollution (gaseous hydrochloric acid) given off by the Leblanc process, used to produce soda ash.

Conservation Movement

The modern conservation movement was first manifested in the forests of India, with the practical application of scientific conservation principles. The conservation ethic that began to evolve included three core principles: that the human activity damaged the environment, that there was a civic duty to maintain the environment for future generations, and that scientific, empirically based methods should be applied to ensure this duty was carried out.

Students from the forestry school at Oxford, on a visit to the forests of Saxony in the year 1892.

Sir James Ranald Martin was prominent in promoting this ideology, publishing many medico-topographical reports that demonstrated the scale of damage wrought through large-scale deforestation and desiccation, and lobbying extensively for the institutionalization of forest conservation activities in British India through the establishment of Forest Departments. The Madras Board of Revenue started local conservation efforts in 1842, headed by Alexander Gibson, a professional botanist who systematically adopted a forest conservation program based on scientific principles. This was the first case of state management of forests in the world. Eventually, the government under Governor-General Lord Dalhousie introduced the first permanent and large-scale forest conservation program in the world in 1855, a model that soon spread to other colonies, as well the

United States. In 1860, the Department banned the use shifting cultivation. Dr. Hugh Cleghorn's 1861 manual, *The forests and gardens of South India*, became the definitive work on the subject and was widely used by forest assistants in the subcontinent.

Sir Dietrich Brandis joined the British service in 1856 as superintendent of the teak forests of Pegu division in eastern Burma. During that time Burma's teak forests were controlled by militant Karen tribals. He introduced the "taungya" system, in which Karen villagers provided labour for clearing, planting and weeding teak plantations. He formulated new forest legislation and helped establish research and training institutions. The Imperial Forest School at Dehradun was founded by him.

Formation of Environmental Protection Societies

The late 19th century saw the formation of the first wildlife conservation societies. The zoologist Alfred Newton published a series of investigations into the *Desirability of establishing a 'Close-time' for the preservation of indigenous animals* between 1872 and 1903. His advocacy for legislation to protect animals from hunting during the mating season led to the formation of the Plumage League (later the Royal Society for the Protection of Birds) in 1889. The society acted as a protest group campaigning against the use of great crested grebe and kittiwake skins and feathers in fur clothing. The Society attracted growing support from the suburban middle-classes, and influenced the passage of the Sea Birds Preservation Act in 1869 as the first nature protection law in the world.

For most of the century from 1850 to 1950, however, the primary environmental cause was the mitigation of air pollution. The Coal Smoke Abatement Society was formed in 1898 making it one of the oldest environmental NGOs. It was founded by artist Sir William Blake Richmond, frustrated with the pall cast by coal smoke. Although there were earlier pieces of legislation, the Public Health Act 1875 required all furnaces and fireplaces to consume their own smoke.

John Ruskin an influential thinker who articulated the Romantic ideal of environmental protection and conservation.

Systematic and general efforts on behalf of the environment only began in the late 19th century; it grew out of the amenity movement in Britain in the 1870s, which was a reaction to industrialization, the growth of cities, and worsening air and water pollution. Starting with the formation of the

Commons Preservation Society in 1865, the movement championed rural preservation against the encroachments of industrialisation. Robert Hunter, solicitor for the society, worked with Hardwicke Rawnsley, Octavia Hill, and John Ruskin to lead a successful campaign to prevent the construction of railways to carry slate from the quarries, which would have ruined the unspoilt valleys of Newlands and Ennerdale. This success led to the formation of the Lake District Defence Society (later to become The Friends of the Lake District).

In 1893 Hill, Hunter and Rawnsley agreed to set up a national body to coordinate environmental conservation efforts across the country; the "National Trust for Places of Historic Interest or Natural Beauty" was formally inaugurated in 1894. The organisation obtained secure footing through the 1907 National Trust Bill, which gave the trust the status of a statutory corporation. and the bill was passed in August 1907.

An early "Back-to-Nature" movement, which anticipated the romantic ideal of modern environmentalism, was advocated by intellectuals such as John Ruskin, William Morris, and Edward Carpenter, who were all against consumerism, pollution and other activities that were harmful to the natural world. The movement was a reaction to the urban conditions of the industrial towns, where sanitation was awful, pollution levels intolerable and housing terribly cramped. Idealists championed the rural life as a mythical Utopia and advocated a return to it. John Ruskin argued that people should return to a *small piece of English ground, beautiful, peaceful, and fruitful. We will have no steam engines upon it . . . we will have plenty of flowers and vegetables . . . we will have some music and poetry; the children will learn to dance to it and sing it.*

Practical ventures in the establishment of small cooperative farms were even attempted and old rural traditions, without the "taint of manufacture or the canker of artificiality", were enthusiastically revived, including the Morris dance and the maypole.

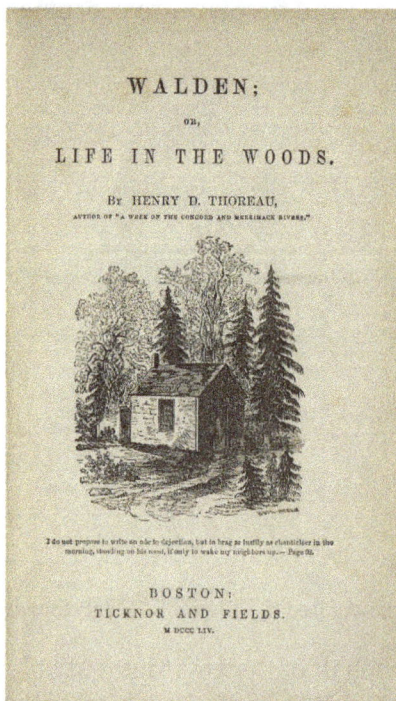

Original title page of *Walden* by Henry David Thoreau.

The movement in the United States began in the late 19th century, out of concerns for protecting the natural resources of the West, with individuals such as John Muir and Henry David Thoreau making key philosophical contributions. Thoreau was interested in peoples' relationship with nature and studied this by living close to nature in a simple life. He published his experiences in the book *Walden*, which argues that people should become intimately close with nature. Muir came to believe in nature's inherent right, especially after spending time hiking in Yosemite Valley and studying both the ecology and geology. He successfully lobbied congress to form Yosemite National Park and went on to set up the Sierra Club in 1892. The conservationist principles as well as the belief in an inherent right of nature were to become the bedrock of modern environmentalism. However, the early movement in the U.S. developed with a contradiction; preservationists like John Muir wanted land and nature set aside for its own sake, and conservationists, such as Gifford Pinchot (appointed as the first Chief of the US Forest Service from 1905-1910), wanted to manage natural resources for human use.

20th Century

In the 20th century, environmental ideas continued to grow in popularity and recognition. Efforts were starting to be made to save some wildlife, particularly the American bison. The death of the last passenger pigeon as well as the endangerment of the American bison helped to focus the minds of conservationists and popularize their concerns. In 1916 the National Park Service was founded by US President Woodrow Wilson. Pioneers of the movement called for more efficient and professional management of natural resources. They fought for reform because they believed the destruction of forests, fertile soil, minerals, wildlife and water resources would lead to the downfall of society. The group that has been the most active in recent years is the climate movement.

> *The conservation of natural resources is the fundamental problem. Unless we solve that problem, it will avail us little to solve all others.*
>
> *Theodore Roosevelt (4 October 1907)*

The U.S movement did not really take off until after World War II as people began to recognize the costs of environmental negligence, disease, and widespread air and water pollution through the occurrence of several environmental disasters that occurred post-World War II. Aldo Leopold wrote "A Sand County Almanac" in the 1940s. He believed in a land ethic that recognized that maintaining the "beauty, integrity, and health of natural systems" as a moral and ethical imperative.

Another important book in the promotion of the environmental movement was Rachel Carson's "Silent Spring" about declining bird populations due to DDT, an insecticide, pollution and man's attempts to control nature through use of synthetic substances. Both of these books helped bring the issues into the public eye Rachel Carson's "Silent Spring" sold over two million copies.

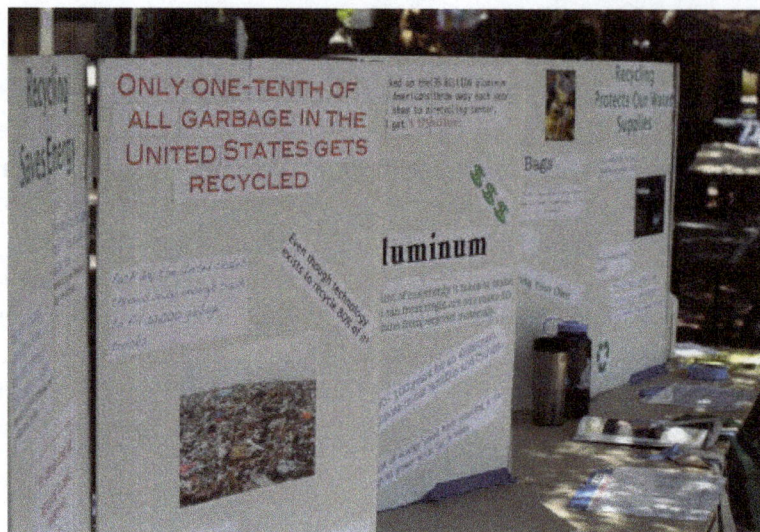

Earth Day 2007 at City College, San Diego

The first Earth Day was celebrated on 22 April 1970. Its founder, former Wisconsin Senator, Gaylord Nelson was inspired to create this day of environmental education and awareness after seeing the oil spill off the coast of Santa Barbara in 1969. Greenpeace was created in 1971 as an organization that believed that political advocacy and legislation were ineffective or inefficient solutions and supported non-violent action. 1980 saw the creation of Earth First!, a group with an ecocentric view of the world --- believing in equality between the rights of humans to flourish, the rights of all other species to flourish and the rights of life-sustaining systems to flourish.

In the 1950s, 1960s, and 1970s, several events illustrated the magnitude of environmental damage caused by humans. In 1954, a hydrogen bomb test at Bikini Atoll exposed the 23 man crew of the Japanese fishing vessel *Lucky Dragon 5* to radioactive fallout. In 1967 the oil tanker *Torrey Canyon* ran aground off the coast of Cornwall, and in 1969 oil spilled from an offshore well in California's Santa Barbara Channel. In 1971, the conclusion of a lawsuit in Japan drew international attention to the effects of decades of mercury poisoning on the people of Minamata.

At the same time, emerging scientific research drew new attention to existing and hypothetical threats to the environment and humanity. Among them were Paul R. Ehrlich, whose book *The Population Bomb* (1968) revived Malthusian concerns about the impact of exponential population growth. Biologist Barry Commoner generated a debate about growth, affluence and "flawed technology." Additionally, an association of scientists and political leaders known as the Club of Rome published their report *The Limits to Growth* in 1972, and drew attention to the growing pressure on natural resources from human activities.

Meanwhile, technological accomplishments such as nuclear proliferation and photos of the Earth from outer space provided both new insights and new reasons for concern over Earth's seemingly small and unique place in the universe.

In 1972, the United Nations Conference on the Human Environment was held in Stockholm, and for the first time united the representatives of multiple governments in discussion relating to the state of the global environment. This conference led directly to the creation of government environmental agencies and the UN Environment Program.

By the mid-1970s anti-nuclear activism had moved beyond local protests and politics to gain a wider appeal and influence. Although it lacked a single co-ordinating organization the anti-nuclear movement's efforts gained a great deal of attention, especially in the United Kingdom and United States. In the aftermath of the Three Mile Island accident in 1979, many mass demonstrations took place. The largest one was held in New York City in September 1979 and involved 200,000 people.

Since the 1970s, public awareness, environmental sciences, ecology, and technology have advanced to include modern focus points like ozone depletion, global climate change, acid rain, and the potentially harmful genetically modified organisms (GMOs).

United States

Beginning in the conservation movement at the beginning of the 20th century, the contemporary environmental movement's roots can be traced back to Murray Bookchin's *Our Synthetic Environment*, Paul R. Ehrlich's The Population Bomb, and Rachel Carson's *Silent Spring*. American environmentalists have campaigned against nuclear weapons and nuclear power in 1960s and 1970s, acid rain in the 1980s, ozone depletion and deforestation in the 1990s, and most recently climate change and global warming.

The United States passed many pieces of environmental legislation in the 1970s, such as the Clean Water Act, the Clean Air Act, the Endangered Species Act, and the National Environmental Policy Act. These remain as the foundations for current environmental standards.

Timeline of US Environmental History

- 1832- Hot Springs Reservation
- 1864- Yosemite Valley
- 1872- Yellowstone National Park
- 1892- Sierra Club
- 1916- National Park Service Organic Act
- 1916- National Audubon Society
- 1949- UN Scientific Conference on the Conservation and Utilization of Resources
- 1961- World Wildlife Foundation
- 1964- Land and Water Conservation Act
- 1964- National Wilderness Preservation System
- 1968- National Trails System Act
- 1968- National Wild and Scenic Rivers System/Wild and Scenic Rivers Act

- 1969- National Environmental Policy Act

- 1970- First Earth Day- 22 April

- 1970- Clean Air Act

- 1970- Environmental Protection Agency

- 1971- Greenpeace

- 1972- Clean Water Act

- 1973- Endangered Species Act

- 1980- Earth First!

- 1992- UN Earth Summit in Rio de Janeiro

Latin America

After the International Environmental Conference in Stockholm in 1972 Latin American officials returned with a high hope of growth and protection of the fairly untouched natural resources. Governments spent millions of dollars, and created departments and pollution standards. However, the outcomes have not always been what officials had initially hoped. Activists blame this on growing urban populations and industrial growth. Many Latin American countries have had a large inflow of immigrants that are living in substandard housing. Enforcement of the pollution standards is lax and penalties are minimal; in Venezuela, the largest penalty for violating an environmental law is 50,000 bolivar fine ($3,400) and 3 days in jail. In the 1970s or 1980s many Latin American countries were transitioning from military dictatorships to democratic governments.

Brazil

In 1992, Brazil came under scrutiny with the United Nations Conference on Environment and Development in Rio de Janeiro. Brazil has a history of little environmental awareness. It has the highest biodiversity in the world and also the highest amount of habitat destruction. One-third of the world's forests lie in Brazil, and they have the largest river, The Amazon, and the largest rainforest, the Amazon Rainforest. The people have raised funds to create state parks and increase the consciousness of people who have destroyed forests and polluted waterways. They have several organizations that have fronted the environmental movement. The Blue Wave Foundation was created in 1989 and has partnered with advertising companies to promote national education campaigns to keep Brazil's beaches clean. Funatura was created in 1986 and is a wildlife sanctuary program. Pro-Natura International is a private environmental organization created in 1986.

Europe

In 1952 the Great London Smog episode killed thousands of people and led the UK to create the first Clean Air Act in 1956. In 1957 the first major nuclear accident occurred in Windscale in northern England. The supertanker *Torrey Canyon* ran aground off the coast of Cornwall in 1967 causing the first major oil leak that killed marine life along the coast. In 1972, in Stockholm, the United

Nations Conference on the Human Environment created the UN Environment Programme. The EU's environmental policy was formally founded by a European Council declaration and the first five-year environment programme was adopted. The main idea of the declaration was that prevention is better than the cure and the polluter should pay. 1979 saw the partial meltdown of Three Mile Island in the USA.

In the 1980s the green parties that were created a decade before began to have some political success.. In 1986, there was a nuclear accident in Chernobyl, Ukraine. The end of the 1980s and start of the 1990s saw the fall of communism across central and Eastern Europe, the fall of the [Berlin Wall], and the Union of East and West Germany. In 1992 there was a UN summit held in Rio de Janeiro where Agenda 21 was adopted. The Kyoto Protocol was created in 1997 which set specific targets and deadlines to reduce global greenhouse gas emissions. In the early 2000s activists believed that environmental policy concerns were overshadowed by energy security, globalism, and terrorism.

Asia

Middle East

The environmental movement is reaching the less developed world with different degrees of success. The Arab world, including the Middle East and North Africa, has different adaptations of the environmental movement. Countries on the Persian Gulf have high incomes and rely heavily on the large amount of energy resources in the area. Each country in the Arab world has varying combinations of low or high amounts of natural resources and low or high amounts of labor.

The League of Arab States has one specialized sub-committee, of 12 standing specialized sub-committee in the Foreign Affairs Ministerial Committees, which deals with Environmental Issues. Countries in the League of Arab States have demonstrated an interest in environmental issue, on paper some environmental activists have doubts about the level of commitment to environmental issues;; being a part of the world community may have obliged these countries to portray concern for the environment. Initial level of environmental awareness may be the creation of a ministry of the environment. The year of establishment of a ministry is also indicative of level of engagement. Saudi Arabia was the first to establish environmental law in 1992 followed by Egypt in 1994. Somalia is the only country without environmental law. In 2010 the Environmental Performance Index listed Algeria as the top Arab country at 42 of 163; Morocco was at 52 and Syria at 56. The Environmental Performance Index measures the ability of a country to actively manage and protect their environment and the health of their citizens. A weighted index is created by giving 50% weight for environmental health objective (health) and 50% for ecosystem vitality (ecosystem); values range from 0-100. No Arab countries were in the top quartile, and 7 countries were in the lowest quartile.

South Korea and Taiwan

South Korea and Taiwan experienced similar growth in industrialization from 1965-1990 with few environmental controls. South Korea's Han River and Nakdong River were so polluted by unchecked dumping of industrial waste that they were close to being classified as biologically dead.

Taiwan's formula for balanced growth was to prevent industrial concentration and encourage manufacturers to set up in the countryside. This led to 20% of the farmland being polluted by industrial waste and 30% of the rice grown on the island was contaminated with heavy metals. Both countries had spontaneous environmental movements drawing participants from different classes. Their demands were linked with issues of employment, occupational health, and agricultural crisis. They were also quite militant; the people learned that protesting can bring results. The polluting factories were forced to make immediate improvements of the conditions or pay compensation to victims. Some were even forced to shut down or move locations. The people were able to force the government to come out with new restrictive rules on toxins, industrial waste, and air pollution. All of these new regulations caused the migration of those polluting industries from Taiwan and South Korea to China and other countries in Southeast Asia with more relaxed environmental laws.

China

China's environmental movement is characterized by spontaneous alliances that often only occur at the local level. The Chinese have realized the ability of riots and protests to have success and had led to an increase in disputes in China by 30% since 2005 to more than 50,000 events. Protests cover topics such as environmental issues, land-loss, income, and political issues. They have also grown in size from about 10 people or fewer in the mid-1990s to 52 people per incident in 2004. China has more relaxed environmental laws than other countries in Asia, so many polluting factories have relocated to China causing pollution in China. Water pollution, water scarcity, soil pollution, soil degradation, and desertification are issues currently in discussion in China. The groundwater table of the North China Plain is dropping by 1.5 m (5 ft) per year. This groundwater table occurs in the region of China that produces 40% of the country's grain.

India

Environmental and public health is an ongoing struggle within India. The first seed of an environmental movement in India was the foundation in 1964 of *Dasholi Gram Swarajya Sangh*, a labour coperative started by Chandi Prasad Bhatt. It was inaugurated by Sucheta Kriplani and founded on a land donated by Shyma Devi. This initiative was eventually followed up with the Chipko movement starting in 1974.

The most severe single event underpinning the movement was the Bhopal gas leakage on 3 December 1984. 40 tons of methyl isocyanate was released, immediately killing 2,259 people and ultimately affecting 700,000 citizens.

India has a national campaign against Coca Cola and Pepsi Cola plants due to their practices of drawing ground water and contaminating fields with sludge. The movement is characterized by local struggles against intensive aquaculture farms. The most influential part of the environmental movement in India is the anti-dam movement. Dam creation has been thought of as a way for India to catch up with the West by connecting to the power grid with giant dams, coal or oil-powered plants, or nuclear plants. Jhola Aandolan a mass movement is conducting as fighting against polyethylene carry bags uses and promoting cloth/jute/paper carry bags to protect environment & nature. Activists in the Indian environmental movement consider global warming, sea levels rising, and glaciers retreating decreasing the amount of water flowing into streams to be the biggest challenges for them to face in the early twenty-first century.

Bangladesh

Mithun Roy Chowdhury, President, Save Nature & Wildlife (SNW), Bangladesh, insisted that the people of Bangladesh raise their voice against Tipaimukh Dam, being constructed by the Government of India. He said Tipaimukh Dam project will be another "death trap for Bangladesh like the Farakka Barrage," that would lead to an environmental disaster for 50 million people in the Meghna River basin. He said that this project will start desertification in Bangladesh.

Scope of the Movement

Before flue-gas desulfurization was installed, the air-polluting emissions from this power plant in New Mexico contained excessive amounts of sulfur dioxide.

Environmental science is the study of the interactions among the physical, chemical and biological components of the environment.

- Ecology, or ecological science, is the scientific study of the distribution and abundance of living organisms and how these properties are affected by interactions between the organisms and their environment.

Primary Focus Points

The environmental movement is broad in scope and can include any topic related to the environment, conservation, and biology, as well as preservation of landscapes, flora, and fauna for a variety of purposes and uses. See List of environmental issues. When an act of violence is committed against someone or some institution in the name of environmental defense it is referred to as eco terrorism.

- The conservation movement seeks to protect natural areas for sustainable consumption, as well as traditional (hunting, fishing, trapping) and spiritual use.

- Environmental conservation is the process in which one is involved in conserving the natural aspects of the environment. Whether through reforestation, recycling, or pollution control, environmental conservation sustains the natural quality of life.

- Environmental health movement dates at least to Progressive Era, and focuses on urban standards like clean water, efficient sewage handling, and stable population growth. Environmental health could also deal with nutrition, preventive medicine, aging, and other concerns specific to human well-being. Environmental health is also seen as an indicator for the state of the environment, or an early warning system for what may happen to humans

- Environmental justice is a movement that began in the U.S. in the 1980s and seeks an end to environmental racism and prevent low-income and minority communities from an unbalanced exposure to highways, garbage dumps, and factories. The Environmental Justice movement seeks to link "social" and "ecological" environmental concerns, while at the same time preventing de facto racism, and classism. This makes it particularly adequate for the construction of labor-environmental alliances.

- Ecology movement could involve the Gaia Theory, as well as Value of Earth and other interactions between humans, science, and responsibility.

- Bright green environmentalism is a currently popular sub-movement, which emphasizes the idea that through technology, good design and more thoughtful use of energy and resources, people can live responsible, sustainable lives while enjoying prosperity.

- Light green, and dark green environmentalism are yet other sub-movements, respectively distinguished by seeing environmentalism as a lifestyle choice (light greens), and promoting reduction in human numbers and/or a relinquishment of technology (dark greens)

- Deep Ecology is an ideological spinoff of the ecology movement that views the diversity and integrity of the planetary ecosystem, in and for itself, as its primary value.

- The anti-nuclear movement opposes the use of various nuclear technologies. The initial anti-nuclear objective was nuclear disarmament and later the focus began to shift to other issues, mainly opposition to the use of nuclear power. There have been many large anti-nuclear demonstrations and protests. Major anti-nuclear groups include Campaign for Nuclear Disarmament, Friends of the Earth, Greenpeace, International Physicians for the Prevention of Nuclear War, and the Nuclear Information and Resource Service.

Environmental Law and Theory

Property Rights

Many environmental lawsuits question the legal rights of property owners, and whether the general public has a right to intervene with detrimental practices occurring on someone else's land. Environmental law organizations exist all across the world, such as the Environmental Law and Policy Center in the midwestern United States.

Citizens' Rights

One of the earliest lawsuits to establish that citizens may sue for environmental and aesthetic harms was Scenic Hudson Preservation Conference v. Federal Power Commission, decided in 1965 by the Second Circuit Court of Appeals. The case helped halt the construction of a power plant on Storm King Mountain in New York State.

Nature's Rights

Christopher D. Stone's 1972 essay, "Should trees have standing?" addressed the question of whether natural objects themselves should have legal rights. In the essay, Stone suggests that his argument is valid because many current rightsholders (women, children) were once seen as objects.

Environmental Reactivism

Numerous criticisms and ethical ambiguities have led to growing concerns about technology, including the use of potentially harmful pesticides, water additives like fluoride, and the extremely dangerous ethanol-processing plants.

NIMBY syndrome refers to public outcry caused by knee-jerk reaction to an unwillingness to be exposed to even necessary developments. Some serious biologists and ecologists created the scientific ecology movement which would not confuse empirical data with visions of a desirable future world.

Environmentalism Today

Composite images of Earth generated by NASA in 2001 (left) and 2002 (right).

Today, the sciences of ecology and environmental science, in addition to any aesthetic goals, provide the basis of unity to some of the serious environmentalists. As more information is gathered in scientific fields, more scientific issues like biodiversity, as opposed to mere aesthetics, are a concern to environmentalists. Conservation biology is a rapidly developing field.

In recent years, the environmental movement has increasingly focused on global warming as one of the top issues. As concerns about climate change moved more into the mainstream, from the connections drawn between global warming and Hurricane Katrina to Al Gore's film An Inconvenient Truth, more and more environmental groups refocused their efforts. In the United States, 2007 witnessed the largest grassroots environmental demonstration in years, Step It Up 2007, with rallies in over 1,400 communities and all 50 states for real global warming solutions.

Many religious organizations and individual churches now have programs and activities dedicated to environmental issues. The religious movement is often supported by interpretation of scriptures. Most major religious groups are represented including Jewish, Islamic, Anglican, Orthodox, Evangelical, Christian and Catholic.

Radical Environmentalism

Radical environmentalism emerged from an ecocentrism-based frustration with the co-option of mainstream environmentalism. The radical environmental movement aspires to what scholar Christopher Manes calls "a new kind of environmental activism: iconoclastic, uncompromising, discontented with traditional conservation policy, at times illegal ..." Radical environmentalism presupposes a need to reconsider Western ideas of religion and philosophy (including capitalism, patriarchy and globalization) sometimes through "resacralising" and reconnecting with nature. Greenpeace represents an organisation with a radical approach, but has contributed in serious ways towards understanding of critical issues, and has a science-oriented core with radicalism as a means to media exposure. Groups like Earth First! take a much more radical posture. Some radical environmentalist groups, like Earth First! and the Earth Liberation Front, illegally sabotage or destroy infrastructural capital.

Criticisms

Conservative critics of the movement characterize it as radical and misguided. Especially critics of the United States Endangered Species Act, which has come under scrutiny lately, and the Clean Air Act, which they said conflict with private property rights, corporate profits and the nation's overall economic growth. Critics also challenge scientific evidence for global warming. They argue that the Environmental Movement has diverted attention from more pressing issues.

Environmental Impact Assessment

See Economic impact analysis for an Economic Impact Assessment. See Environmental impact statement for the report/document produced by the process of Environmental Impact Assessment.

Environmental assessment' (EA) is the term used for the assessment of the environmental consequences (positive and negative) of a plan, policy, program, or project prior to the decision to move forward with the proposed action. In this context, the term 'environmental impact assessment' (EIA) is usually used when applied to concrete projects and the term 'strategic environmental assessment' applies to policies, plans and programmes (Fischer, 2016). Environmental assessments may be governed by rules of administrative procedure regarding public participation and documentation of decision making, and may be subject to judicial review.

The purpose of the assessment is to ensure that decision makers consider the environmental impacts when deciding whether or not to proceed with a project. The International Association for Impact Assessment (IAIA) defines an environmental impact assessment as "the process of identifying, predicting, evaluating and mitigating the biophysical, social, and other relevant effects of development proposals prior to major decisions being taken and commitments made." EIAs are unique in that they do not require adherence to a predetermined environmental outcome, but rather they require decision makers to account for environmental values in their decisions and to justify those decisions in light of detailed environmental studies and public comments on the potential environmental impacts.

Engineering and consulting companies work hand in hand as contractors for mining, energy, oil & gas companies executing EIAs. Companies operating globally such as Arcadis, Royal HaskoningDHV, Golder Associates, Amec Foster Wheeler, Schlumberger Water Services (a Schlumberger company), ERM are an example of a much bigger pool of expertise globally. These contractors are the ones not only in charge of preparing an EIA study but most importantly getting these studies approved by each country government offices prior to the execution of a project. Each country will also have its own local contractors offering the same kind of service hence breaking out monopolies by increasing the supply of EIAs execution consultants.

History

Environmental impact assessments commenced in the 1960s, as part of increasing environmental awareness. EIAs involved a technical evaluation intended to contribute to more objective decision making. In the United States, environmental impact assessments obtained formal status in 1969, with enactment of the National Environmental Policy Act. EIAs have been used increasingly around the world. The number of "Environmental Assessments" filed every year "has vastly overtaken the number of more rigorous Environmental Impact Statements (EIS)." An Environmental Assessment is a "mini-EIS designed to provide sufficient information to allow the agency to decide whether the preparation of a full-blown Environmental Impact Statement (EIS) is necessary." EIA is an activity that is done to find out the impact that would be done before development will occur.

Methods

General and industry specific assessment methods are available including:

- *Industrial products* - Product environmental life cycle analysis (LCA) is used for identifying and measuring the impact of industrial products on the environment. These EIAs consider activities related to extraction of raw materials, ancillary materials, equipment; production, use, disposal and ancillary equipment.

- *Genetically modified plants* - Specific methods available to perform EIAs of genetically modified organisms include GMP-RAM and INOVA.

- *Fuzzy logic* - EIA methods need measurement data to estimate values of impact indicators. However, many of the environment impacts cannot be quantified, e.g. landscape quality, lifestyle quality and social acceptance. Instead information from similar EIAs, expert judgment and community sentiment are employed. Approximate reasoning methods known as fuzzy logic can be used. A fuzzy arithmetic approach has also been proposed and implemented using a software tool (TDEIA). More information can be found at ARAI web site.

Follow-Up

At the end of the project, an audit evaluates the accuracy of the EIA by comparing actual to predicted impacts. The objective is to make future EIAs more valid and effective. Two primary considerations are:

- *Scientific* - to examine the accuracy of predictions and explain errors

- *Management* - to assess the success of mitigation in reducing impacts

Audits can be performed either as a rigorous assessment of the null hypothesis or with a simpler approach comparing what actually occurred against the predictions in the EIA document.

After an EIA, the precautionary and polluter pays principles may be applied to decide whether to reject, modify or require strict liability or insurance coverage to a project, based on predicted harms.

The Hydropower Sustainability Assessment Protocol is a sector specific method for checking the quality of Environmental and Social assessments and management plans.

Around the World

Australia

The history of EIA in Australia could be linked to the enactment of the U.S. National Environment Policy Act (NEPA) in 1970, which made the preparation of environmental impact statements a requirement. In Australia, one might say that the EIA procedures were introduced at a State Level prior to that of the Commonwealth (Federal), with a majority of the states having divergent views to the Commonwealth. One of the pioneering states was New South Wales, whose State Pollution Control Commission issued EIA guidelines in 1974. At a Commonwealth (Federal) level, this was followed by passing of the Environment Protection (Impact of Proposals) Act in 1974. The Environment Protection and Biodiversity Conservation Act 1999 (EPBC) superseded the Environment Protection (Impact of Proposals) Act 1974 and is the current central piece for EIA in Australia on a Commonwealth (Federal) level. An important point to note is that this Commonwealth Act does not affect the validity of the States and Territories environmental and development assessments and approvals; rather the EPBC runs as a parallel to the State/Territory Systems. Overlap between federal and state requirements is addressed via bilateral agreements or one off accreditation of state processes, as provided for in the EPBC Act.

The Commonwealth Level

The EPBC Act provides a legal framework to protect and manage nationally and internationally important flora, fauna, ecological communities and heritage places-defined in the EPBC Act as matters of 'national environmental significance'. Following are the eight matters of 'national environmental significance' to which the EPBC ACT applies:

- World Heritage sites

- National Heritage places

- RAMSAR wetlands of international significance

- Listed threatened species and ecological communities

- Migratory species protected under international agreements

- The Commonwealth marine environment

- Nuclear actions (including uranium mining)

- National Heritage.

In addition to this, the EPBC Act aims at providing a streamlined national assessment and approval process for activities. These activities could be by the Commonwealth, or its agents, anywhere in the world or activities on Commonwealth land; and activities that are listed as having a 'significant impact' on matters of 'national environment significance'.

The EPBC Act comes into play when a person (a 'proponent') wants an action (often called a 'proposal' or 'project') assessed for environmental impacts under the EPBC Act, he or she must refer the project to the Department of Environment, Water, Heritage and the Arts (Australia). This 'referral' is then released to the public, as well as relevant state, territory and Commonwealth ministers, for comment on whether the project is likely to have a significant impact on matters of national environmental significance. The Department of Environment, Water, Heritage and the Arts assess the process and makes recommendation to the minister or the delegate for the feasibility. The final discretion on the decision remains of the minister, which is not solely based on matters of 'national environmental significance' but also the consideration of social and economic impact of the project.

The Australian Government environment minister cannot intervene in a proposal if it has no significant impact on one of the eight matters of 'national environmental significance' despite the fact that there may be other undesirable environmental impacts. This is primarily due to the division of powers between the States and the Federal government and due to which the Australian Government environment minister cannot overturn a state decision.

There are strict civil and criminal penalties for the breach of EPBC Act. Depending on the kind of breach, civil penalty (maximum) may go up to $550,000 for an individual and $5.5 million for a body corporate, or for criminal penalty (maximum) of seven years imprisonment and/or penalty of $46,200.

The State and Territory Level

Australian Capital Territory (ACT)

EIA provisions within Ministerial Authorities in the ACT are found in the Chapters 7 and 8 of the *Planning and Development Act 2007* (ACT). EIA in ACT was previously administered with the help of Part 4 of the Land (Planning and Environment) Act 1991 (Land Act) and Territory Plan (plan for land-use). Note that some EIA may occur in the ACT on Commonwealth land under the EPBC Act (Cth). Further provisions of the *Australian Capital Territory (Planning and Land Management) Act 1988* (Cth) may also be applicable particularly to national land and "designated areas".

New South Wales (NSW)

In New South Wales, the Environment Planning Assessment Act 1979 (EPA) establishes three pathways for EIA. The first is under Part 5.1 of the EPAA, which provides for EIA of 'State Significant Infrastructure' projects. (From June 2011, this Part replaced Part 3A, which previously covered EIA of major projects). The second is under Part 4 of the Act dealing with development control. If a project does not require approval under Part 3A or Part 4 it is then potentially captured by the third pathway, Part 5 dealing with environment impact assessment.

Northern Territory (NT)

The EIA process in Northern Territory is chiefly administered under the Environmental Assessment Act (EAA). Although EAA is the primary tool for EIA in Northern Territory, there are further provisions for proposals in the Inquiries Act 1985 (NT).

Queensland (QLD)

There are four main EIA processes in Queensland. Firstly, under the Integrated Planning Act 1997 (IPA) for development projects other than mining. Secondly, under the Environmental Protection Act 1994 (EP Act) for some mining and petroleum activities. Thirdly, under the State Development and Public Works Organization Act 1971 (State Development Act) for 'significant projects'. Finally, Environment Protection and Biodiversity Conservation Act 1999 (Cth) for 'controlled actions'.

South Australia (SA)

The local governing tool for EIA in South Australia is the Development Act 1993. There are three levels of assessment possible under the Act in the form of an environment impact statement (EIS), a public environmental report (PER) or a Development Report (DR).

Tasmania (TAS)

In Tasmania, an integrated system of legislation is used to govern development and approval process, this system is a mixture of the Environmental Management and Pollution Control Act 1994 (EMPCA), Land Use Planning and Approvals Act 1993 (LUPAA), State Policies and Projects Act 1993 (SPPA), and Resource Management and Planning Appeals Tribunal Act 1993.

Victoria (VIC)

The EIA process in Victoria is intertwined with the Environment Effects Act 1978 and the Ministerial Guidelines for Assessment of Environmental Effects (made under the s. 10 of the EE Act).

Western Australia (WA)

The Environmental Protection Act 1986 (Part 4) provides the legislative framework for the EIA process in Western Australia. The EPA Act oversees the planning and development proposals and assesses their likely impacts on the environment.

Canada

In *Friends of the Oldman River Society v. Canada (Minister of Transportation)*,(SCC 1992) La Forest J of the Supreme Court of Canada described environmental impact assessment in terms of the proper scope of federal jurisdiction with respect to environments matters,

"Environmental impact assessment is, in its simplest form, a planning tool that is now generally regarded as an integral component of sound decision-making."

Supreme Court Justice La Forest cited (Cotton, Emond & 1981 245), "The basic concepts behind environmental assessment are simply stated: (1) early identification and evaluation of all potential environmental consequences of a proposed undertaking; (2) decision making that both guarantees the adequacy of this process and reconciles, to the greatest extent possible, the proponent's development desires with environmental protection and preservation."

La Forest referred to (Jeffrey 1989, 1.2,1.4) and (Emond 1978, p. 5) who described "...environmental assessments as a planning tool with both an information-gathering and a decision-making component" that provide "...an objective basis for granting or denying approval for a proposed development."

Justice La Forest addressed his concerns about the implications of Bill C-45 regarding public navigation rights on lakes and rivers that would contradict previous cases.(La Forest & 1973 178-80)

The Canadian Environmental Assessment Act 2012 (CEAA 2012) "and its regulations establish the legislative basis for the federal practice of environmental assessment in most regions of Canada." CEAA 2012 came into force July 6, 2012 and replaces the former *Canadian Environmental Assessment Act* (1995). EA is defined as a planning tool to identify, understand, assess and mitigate, where possible, the environmental effects of a project.

"The purposes of this Act are: (a) to protect the components of the environment that are within the legislative authority of Parliament from significant adverse environmental effects caused by a designated project; (b) to ensure that designated projects that require the exercise of a power or performance of a duty or function by a federal authority under any Act of Parliament other than this Act to be carried out, are considered in a careful and precautionary manner to avoid significant adverse environmental effects; (c) to promote cooperation and coordinated action between federal and provincial governments with respect to environmental assessments; (d) to promote communication and cooperation with aboriginal peoples with respect to environmental assessments; (e) to ensure that opportunities are provided for meaningful public participation during an environmental assessment; (f) to ensure that an environmental assessment is completed in a timely manner; (g) to ensure that projects, as defined in section 66, that are to be carried out on federal lands, or those that are outside Canada and that are to be carried out or financially supported by a federal authority, are considered in a careful and precautionary manner to avoid significant adverse environmental effects; (h) to encourage federal authorities to take actions that promote sustainable development in order to achieve or maintain a healthy environment and a healthy economy; and (i) to encourage the study of the cumulative effects of physical activities in a region and the consideration of those study results in environmental assessments."

Opposition

Environmental Lawyer Dianne Saxe argued that the CEAA 2012 "allows the federal government to create mandatory timelines for assessments of even the largest and most important projects, regardless of public opposition." (Saxe 2012)

> "Now that federal environmental assessments are gone, the federal government will only assess very large, very important projects. But it's going to do them in a hurry."
>
> *Dianne Saxe*

On 3 August 2012 the Canadian Environmental Assessment Agency nine "designated projects" with their timelines: Enbridge Northern Gateway Pipeline Joint Review Panel (JRP) 18 months; Marathon Platinum Group Metals and Copper Mine Project (JRP): 13 months; Site C Clean Energy Project (JRP) 8.5 months; Deep Geologic Repository Project (JRP) 17 months; Enbridge Northern Gateway Project (JRP) 18 months; Jackpine Mine Expansion Project (JRP) 11.5 months; Pierre River Mine Project: 8 months; New Prosperity Gold-Copper Mine Project (JRP) 7.5 months; Frontier Oil Sands Mine Project (JRP)8.5 months; EnCana/Cenovus Shallow Gas Infill Project (JRP) 5 months.

Saxe compares these timelines with environmental assessments for the Mackenzie Valley Pipeline. Thomas R. Berger, Royal Commissioner of the Mackenzie Valley Pipeline Inquiry (9 May 1977), worked extremely hard to ensure that industrial development on Aboriginal people's land resulted in benefits to those indigenous people.

On 22 April 2013, Official Opposition Environment critic Megan Leslie issued a statement claiming that the federal government's recent changes to "fish habitat protection, the Navigable Waters Protection Act and the Canadian Environmental Assessment Act", along with gutting existing laws and making cuts to science and research, "will be disastrous, not only for the environment, but also for Canadians' health and economic prosperity." On 26 September 2012, Leslie argued that with the changes to the Canadian Environmental Assessment Act that came into effect 6 July 2012, "seismic testing, dams, wind farms and power plants" no longer required any federal environmental assessment. She also claimed that because the CEAA 2012—which she claimed was rushed through Parliament—dismantled the CEAA 1995, the Oshawa ethanol plant project would no longer have a full federal environmental assessment. Mr. Peter Kent (Minister of the Environment) explained that the CEAA 2012 "provides for the Government of Canada and the Environmental Assessment Agency to focus on the large and most significant projects that are being proposed across the country." The 2,000 to 3,000-plus smaller screenings that were in effect under CEAA 1995 became the "responsibility of lower levels of government but are still subject to the same strict federal environmental laws." Anne Minh-Thu Quach, MP for Beauharnois—Salaberry, QC, argued that the mammoth budget bill dismantled 50 years of environmental protection without consulting Canadians about the "colossal changes they are making to environmental assessments." She claimed that the federal government is entering into "limited consultations, by invitation only, months after the damage was done."

China

The Environmental Impact Assessment Law (EIA Law) requires that an environmental impact assessment be completed prior to project construction. However, if a developer completely ignores this requirement and builds a project without submitting an environmental impact statement, the only penalty is that the environmental protection bureau (EPB) may require the developer to do a make-up environmental assessment. If the developer does not complete this make-up assessment within the designated time, only then is the EPB authorized to fine the developer. Even so, the possible fine is capped at a maximum of about US$25,000, a fraction of the overall cost of most major projects. The lack of more stringent enforcement mechanisms has resulted in a significant percentage of projects not completing legally required environmental impact assessments prior to construction.

China's State Environmental Protection Administration (SEPA) used the legislation to halt 30 projects in 2004, including three hydro-power plants under the Three Gorges Project Company. Although one month later (Note as a point of reference, that the typical EIA for a major project in the USA takes one to two years.), most of the 30 halted projects resumed their construction, reportedly having passed the environmental assessment, the fact that these key projects' construction was ever suspended was notable.

A joint investigation by SEPA and the Ministry of Land and Resources in 2004 showed that 30-40% of the mining construction projects went through the procedure of environment impact assessment as required, while in some areas only 6-7% did so. This partly explains why China has witnessed so many mining accidents in recent years.

SEPA alone cannot guarantee the full enforcement of environmental laws and regulations, observed Professor Wang Canfa, director of the centre to help environmental victims at China University of Political Science and Law. In fact, according to Wang, the rate of China's environmental laws and regulations that are actually enforced is estimated at barely 10%.

Egypt

Environmental Impact Assessment (EIA) EIA is implemented in Egypt under the umbrella of the Ministry of state for environmental affairs. The Egyptian Environmental Affairs Agency (EEAA) is responsible for the EIA services.

In June 1997, the responsibility of Egypt's first full-time Minister of State for Environmental Affairs was assigned as stated in the Presidential Decree no.275/1997. From thereon, the new ministry has focused, in close collaboration with the national and international development partners, on defining environmental policies, setting priorities and implementing initiatives within a context of sustainable development.

According to the Law 4/1994 for the Protection of the Environment, the Egyptian Environmental

Affairs Agency (EEAA) was restructured with the new mandate to substitute the institution initially established in 1982. At the central level, EEAA represents the executive arm of the Ministry.

The purpose of EIA is to ensure the protection and conservation of the environment and natural resources including human health aspects against uncontrolled development. The long-term objective is to ensure a sustainable economic development that meets present needs without compromising future generations ability to meet their own needs. EIA is an important tool in the integrated environmental management approach.

EIA must be performed for new establishments or projects and for expansions or renovations of existing establishments according to the Law for the Environment.

EU

There is a wide range of instruments in the Environmental policy of the European Union. Among them the European Union has established a mix of mandatory and discretionary procedures to assess environmental impacts. European Union Directive (85/337/EEC) on Environmental Impact Assessments (known as the *EIA Directive*) was first introduced in 1985 and was amended in 1997. The directive was amended again in 2003, following EU signature of the 1998 Aarhus Convention, and once more in 2009. The initial Directive of 1985 and its three amendments have been codified in Directive 2011/92/EU of 13 December 2011. In 2001, the issue was enlarged to the assessment of plans and programmes by the so-called *Strategic Environmental Assessment (SEA) Directive* (2001/42/EC), which is now in force. Under the EU directive, an EIA must provide certain information to comply. There are seven key areas that are required:

1. Description of the project

    o   Description of actual project and site description

    o   Break the project down into its key components, i.e. construction, operations, decommissioning

    o   For each component list all of the sources of environmental disturbance

    o   For each component all the inputs and outputs must be listed, e.g., air pollution, noise, hydrology

2. Alternatives that have been considered

    o   Examine alternatives that have been considered

    o   Example: in a biomass power station, will the fuel be sourced locally or nationally?

3. Description of the environment

    o   List of all aspects of the environment that may be affected by the development

    o   Example: populations, fauna, flora, air, soil, water, humans, landscape, cultural heritage

    o   This section is best carried out with the help of local experts, e.g. the RSPB in the

UK

4. Description of the significant effects on the environment

   o   The word significant is crucial here as the definition can vary

   o   'Significant' must be defined

   o   The most frequent method used here is use of the Leopold matrix

   o   The matrix is a tool used in the systematic examination of potential interactions

   o   Example: in a windfarm development a significant impact may be collisions with birds

5. Mitigation

   o   This is where EIA is most useful

   o   Once section 4 is complete, it is obvious where impacts are greatest

   o   Using this information ways to avoid negative impacts should be developed

   o   Best working with the developer with this section as they know the project best

   o   Using the windfarm example again construction could be out of bird nesting seasons

6. Non-technical summary (EIS)

   o   The EIA is in the public domain and be used in the decision making process

   o   It is important that the information is available to the public

   o   This section is a summary that does not include jargon or complicated diagrams

   o   It should be understood by the informed lay-person

7. Lack of know-how/technical difficulties

   o   This section is to advise any areas of weakness in knowledge

   o   It can be used to focus areas of future research

   o   Some developers see the EIA as a starting block for poor environmental management

Annexed Projects

All projects are either classified as Annex 1 or Annex 2 projects. Those lying in Annex 1 are large scale developments such as motorways, chemical works, bridges, powerstations etc. These always require an EIA under the Environmental Impact Assessment Directive (85,337,EEC as amended). Annex 2 projects are smaller in scale than those referred to in Annex 1. Member States must determine whether these project shall be made subject to an assessment subject to a set of criteria set out in Annex 3 of codified Directive 2011/92/EU.

The Netherlands

EIA was implemented in Dutch legislation on September 1, 1987. The categories of projects that require an EIA are summarised in Dutch legislation, the Wet milieubeheer. The use of thresholds for activities makes sure that EIA is obligatory for those activities that may have considerable impacts on the environment.

For projects and plans that fit these criteria, an EIA report is required. The EIA report defines a.o. the proposed initiative, it makes clear the impact of that initiative on the environment and compares this with the impact of possible alternatives with less a negative impact.

Hong Kong

EIA in Hong Kong, since 1998, is regulated by the *Environmental Impact Assessment Ordinance 1997*.

The original proposal to construct the Lok Ma Chau Spur Line overground across the Long Valley failed to get through EIA, and the Kowloon–Canton Railway Corporation had to change its plan and build the railway underground. In April 2011, the EIA of the Hong Kong section of the Hong Kong-Zhuhai-Macau Bridge was found to have breached the ordinance, and was declared unlawful. The appeal by the government was allowed in September 2011. However, it was estimated that this EIA court case had increased the construction cost of the Hong Kong section of the bridge by HK$6.5 billion in money-of-the-day prices.

India

The Ministry of Environment, Forests and Climate Change (MoEFCC) of India has been in a great effort in Environmental Impact Assessment in India. The main laws in action are the Water Act(1974), the Indian Wildlife (Protection) Act (1972), the Air (Prevention and Control of Pollution) Act (1981) and the Environment (Protection) Act (1986),Biological Diversity Act(2002). The responsible body for this is the Central Pollution Control Board. Environmental Impact Assessment (EIA) studies need a significant amount of primary and secondary environmental data. Primary data are those collected in the field to define the status of the environment (like air quality data, water quality data etc.). Secondary data are those collected over the years that can be used to understand the existing environmental scenario of the study area. The environmental impact assessment (EIA) studies are conducted over a short period of time and therefore the understanding of the environmental trends, based on a few months of primary data, has limitations. Ideally, the primary data must be considered along with the secondary data for complete understanding of the existing environmental status of the area. In many EIA studies, the secondary data needs could be as high as 80% of the total data requirement. EIC is the repository of one stop secondary data source for environmental impact assessment in India.

The Environmental Impact Assessment (EIA) experience in India indicates that the lack of timely availability of reliable and authentic environmental data has been a major bottle neck in achieving the full benefits of EIA. The environment being a multi-disciplinary subject, a multitude of agencies are involved in collection of environmental data. However, no single organization in India tracks available data from these agencies and makes it available in one place in a form required by environmental impact assessment practitioners. Further, environmental data is not available in enhanced forms that improve the quality of the EIA. This makes it harder and more time-consuming to generate environmental impact assessments and receive timely environmental clearances from regulators. With this background, the Environmental Information Centre (EIC) has been set up to serve as a professionally managed clearing house of environmental information that can be used by MoEF, project proponents, consultants, NGOs and other stakeholders involved in the process of environmental impact assessment in India. EIC caters to the need of creating and disseminating of organized environmental data for various developmental initiatives all over the country.

EIC stores data in GIS format and makes it available to all environmental impact assessment studies and to EIA stakeholders in a cost effective and timely manner. So that we can manage that in different proportions such as remedy measures etc.,

Korea, South

Recycling culture and policy Ministry of Environment

Malaysia

In Malaysia, Section 34A, Environmental Quality Act, 1974 requires developments that have significant impact to the environment are required to conduct the Environmental impact assessment.

Nepal

In Nepal, EIA has been integrated in major development projects since the early 1980s. In the planning history of Nepal, the sixth plan (1980–85), for the first time, recognized the need for EIA with the establishment of Environmental Impact Study Project (EISP) under the Department of Soil Conservation in 1982 to develop necessary instruments for integration of EIA in infrastructure development projects. However, the government of Nepal enunciated environment conservation related policies in the seventh plan (NPC, 1985–1990). To enforce this policy and make necessary arrangements, a series of guidelines were developed, thereby incorporating the elements of environmental factors right from the project formulation stage of the development plans and projects and to avoid or minimize adverse effects on the ecological system. In addition, it has also emphasized that EIAs of industry, tourism, water resources, transportation, urbanization, agriculture, forest and other developmental projects be conducted.

In Nepal, the government's Environmental Impact Assessment Guideline of 1993 inspired the enactment of the Environment Protection Act (EPA) of 1997 and the Environment Protection Rules (EPR) of 1997 (EPA and EPR have been enforced since 24 and 26 June 1997 respectively in Nepal) to internalizing the environmental assessment system. The process institutionalized the EIA process in development proposals and enactment, which makes the integration of IEE and EIA legally binding to the prescribed projects. The projects, requiring EIA or IEE, are included in Schedules 1 and 2 of the EPR, 1997 (GoN/MoLJPA 1997). Progresses were made in the Environmental protection issue during the 8th five-year plan (1992–1997). The following development in Environmental protection were achieved during that time:

- Formulation of Environmental Protection Act 1997

- Establishment of Ministry of Environment

- Development of National Environmental Policies and Action Plan, EIA guidelines developed

- Consideration of environmental concerns in hydropower projects

- Development of industrial, irrigation and agricultural policies that undertook environmental concerns

New Zealand

In New Zealand, EIA is usually referred to as *Assessment of Environmental Effects* (AEE). The first use of EIA's dates back to a Cabinet minute passed in 1974 called Environmental Protection and Enhancement Procedures. This had no legal force and only related to the activities of government departments. When the Resource Management Act was passed in 1991, an EIA was required as part of a resource consent application. Section 88 of the Act specifies that the AEE must include "such detail as corresponds with the scale and significance of the effects that the activity may have on the environment". While there is no duty to consult any person when making a resource consent application (Sections 36A and Schedule 4), proof of consultation is almost certain required by local councils when they decide whether or not to publicly notify the consent application under Section 93.

Russian Federation

As of 2004, the state authority responsible for conducting the State EIA in Russia has been split between two Federal bodies: 1) Federal service for monitoring the use of natural resources – a part of the Russian Ministry for Natural Resources and Environment and 2) Federal Service for Ecological, Technological and Nuclear Control. The two main pieces of environmental legislation in Russia are: The Federal Law 'On Ecological Expertise, 1995 and the 'Regulations on Assessment of Impact from Intended Business and Other Activity on Environment in the Russian Federation, 2000.

Federal Service for Monitoring the Use of Natural Resources

In 2006, the parliament committee on ecology in conjunction with the Ministry for Natural Resources and Environment, created a working group to prepare a number of amendments to existing legislation to cover such topics as stringent project documentation for building of potentially environmentally damaging objects as well as building of projects on the territory of protected areas. There has been some success in this area, as evidenced from abandonment of plans to construct a gas pipe-line through the only remaining habitat of the critically endangered Amur leopard in the Russian Far East.

Federal Service for Ecological, Technological and Nuclear Control

The government's decision to hand over control over several important procedures, including state EIA in the field of all types of energy projects, to the Federal Service for Ecological, Technological and Nuclear Control had caused a major controversy and criticism from environmental groups that blamed the government for giving nuclear power industry control over the state EIA.

Not surprisingly the main problem concerning State EIA in Russia is the clear differentiation of jurisdiction between the two above-mentioned Federal bodies.

Sri Lanka

One popular approach to assist in smart growth in democratic countries is for law-makers to require prospective developers to prepare environmental impact assessments of their plans as a condition for state and/or local governments to go for Environmental Impact Assessments.

These reports often indicate how significant impacts the development generates can be mitigated, usually at developer expense. These assessments are frequently controversial. Conservationists, neighborhood advocacy groups and NIMBYs are often skeptical about such impact reports, even when prepared by independent agencies and approved by decision makers rather than promoters. Conversely, developers sometimes strongly resist requirements to implement the mitigation measures required by the local government, as they may be quite costly.

The importance of the Environmental Impact Assessment as an effective tool for the purpose of integrating environmental considerations with development planning is highly recognized in Sri Lanka. The application of this technique is considered as a means of ensuring that the likely effects of new development projects on the environment are fully understood and taken into account before development is allowed to proceed. The importance of this management tool to foresee potential environmental impacts and problems caused by proposed projects and its use as a mean to make project more suitable to the environment are highly appreciated.

United States

The National Environmental Policy Act of 1969 (NEPA), enacted in 1970, established a policy of environmental impact assessment for federal agency actions, federally funded activities or federally permitted/licensed activities that in the U. S. is termed "environmental review" or simply "the NEPA process." The law also created the Council on Environmental Quality, which promulgated regulations to codify the law's requirements. Under United States environmental law an Environmental Assessment (EA) is compiled to determine the need for an *Environmental Impact Statement* (EIS). Federal or federalized actions expected to subject or be subject to significant environmental impacts will publish a Notice of Intent to Prepare an EIS as soon as significance is known. Certain actions of federal agencies must be preceded by the NEPA process. Contrary to a widespread misconception, NEPA does not prohibit the federal government or its licensees/permittees from harming the environment, nor does it specify any penalty if an environmental impact assessment turns out to be inaccurate, intentionally or otherwise. NEPA requires that plausible statements as to the prospective impacts be disclosed in advance. The purpose of NEPA process is to ensure that the decision maker is fully informed of the environmental aspects and consequences prior to making the final decision.

Environmental Assessment

An environmental assessment (EA) is an environmental analysis prepared pursuant to the National Environmental Policy Act to determine whether a federal action would significantly affect the environment and thus require a more detailed *Environmental Impact Statement (EIS)*. The certified release of an Environmental Assessment results in either a *Finding of No Significant Impact (FONSI)* or an EIS.

The Council on Environmental Quality (CEQ), which oversees the administration of NEPA, issued regulations for implementing the NEPA in 1979. Eccleston reports that the NEPA regulations barely mention preparation of EAs. This is because the EA was originally intended to be a simple document used in relatively rare instances where an agency was not sure if the potential significance of an action would be sufficient to trigger preparation of an EIS. But today, because EISs are so much longer and complicated to prepare, federal agencies are going to great effort to avoid preparing EISs by using EAs, even in cases where the use of EAs may be inappropriate. The ratio of EAs that are being issued compared to EISs is about 100 to 1.

Likewise, even the preparation of an accurate EA is viewed today as an onerous burden by many entities responsible for the environmental review of a proposal. Federal agencies have responded by streamlining their regulations that implement NEPA environmental review, by defining categories of projects that by their well understood nature may be safely excluded from review under NEPA, and by drawing up lists of project types that have negligible material impact upon the environment and can thus be exempted.

The Environmental Assessment is a concise public document prepared by the federal action agency that serves to:

1. briefly provide sufficient evidence and analysis for determining whether to prepare an EIS or a Finding of No Significant Impact (FONSI)

2. Demonstrate compliance with the act when no EIS is required

3. facilitate the preparation of an EIS when a FONSI cannot be demonstrated

The Environmental Assessment includes a brief discussion of the purpose and need of the proposal and of its alternatives as required by NEPA 102(2)(E), and of the human environmental impacts resulting from and occurring to the proposed actions and alternatives considered practicable, plus a listing of studies conducted and agencies and stakeholders consulted to reach these conclusions. The action agency must approve an EA before it is made available to the public. The EA is made public through notices of availability by local, state, or regional clearing houses, often triggered by the purchase of a public notice advertisement in a newspaper of general circulation in the proposed activity area.

Structure

The structure of a generic Environmental Assessment is as follows:

1. Summary

2. Introduction

  o Background

  o Purpose and Need for Action

  o Proposed Action

  o Decision Framework

  o Public Involvement

  o Issues

3. Alternatives, including the Proposed Action

  o Alternatives

  o Mitigation Common to All Alternatives

  o Comparison of Alternatives

4. Environmental Consequences

5. Consultation and Coordination

Procedure

The EA becomes a draft public document when notice of it is published, usually in a newspaper of general circulation in the area affected by the proposal. There is a 15-day review period required for an Environmental Assessment (30 days if exceptional circumstances) while the document is made available for public commentary, and a similar time for any objection to improper process. Commenting on the Draft EA is typically done in writing or email, submitted to the lead action agency as published in the notice of availability. An EA does not require a public hearing for verbal comments. Following the mandated public comment period, the lead action agency responds to any comments, and certifies either a FONSI or a Notice of Intent (NOI) to prepare an EIS in its public environmental review record. The preparation of an EIS then generates a similar but more lengthy, involved and expensive process.

Environmental Impact Statement

The adequacy of an environmental impact statement (EIS) can be challenged in federal court. Major proposed projects have been blocked because of an agency's failure to prepare an acceptable EIS. One prominent example was the Westway landfill and highway development in and along the Hudson River in New York City. Another prominent case involved the Sierra Club suing the Nevada Department of Transportation over its denial of the club's request to issue a supplemental EIS addressing air emissions of particulate matter and hazardous air pollutants in the case of widening U.S. Route 95 through Las Vegas. The case reached the United States Court of Appeals for the Ninth Circuit, which led to construction on the highway being halted until the court's final decision. The case was settled prior to the court's final decision.

Several state governments that have adopted "little NEPAs," state laws imposing EIS require-ments for particular state actions. Some those state laws such as the California Environmental Quality Act refer to the required environmental impact study as an environmental impact report.

These variety of state requirements are yielding voluminous data not just upon impacts of individ-ual projects, but also to elucidate scientific areas that had not been sufficiently researched. For ex-ample, in a seemingly routine *Environmental Impact Report* for the city of Monterey, California, information came to light that led to the official federal endangered species listing of Hickman's potentilla, a rare coastal wildflower.

Transboundary Application

Environmental threats do not respect national borders. International pollution can have detri-mental effects on the atmosphere, oceans, rivers, aquifers, farmland, the weather and biodiversity. Global climate change is transnational. Specific pollution threats include acid rain, radioactive contamination, debris in outer space, stratospheric ozone depletion and toxic oil spills. The Cher-nobyl disaster, precipitated by a nuclear accident on April 26, 1986, is a stark reminder of the devastating effects of transboundary nuclear pollution.

Environmental protection is inherently a cross-border issue and has led to the creation of trans-national regulation via multilateral and bilateral treaties. The United Nations Conference on the Human Environment (UNCHE or Stockholm Conference) held in Stockholm in 1972 and the Unit-ed Nations Conference on the Environment and Development (UNCED or Rio Summit, Rio Con-ference, or Earth Summit) held in Rio de Janeiro in 1992 were key in the creation of about 1,000 international instruments that include at least some provisions related to the environment and its protection.

The United Nations Economic Commission for Europe's Convention on Environmental Impact Assessment in a Transboundary Context was negotiated to provide an international legal frame-work for transboundary EIA.

However, as there is no universal legislature or administration with a comprehensive mandate, most international treaties exist parallel to one another and are further developed without the benefit of consideration being given to potential conflicts with other agreements. There is also the issue of international enforcement. This has led to duplications and failures, in part due to an inability to enforce agreements. An example is the failure of many international fisheries regimes to restrict harvesting practises.

Criticism

As per Jay *et al.*, EIA is used as a decision aiding tool rather than decision making tool. There is growing dissent about them as their influence on decisions is limited. Improved training for prac-titioners, guidance on best practice and continuing research have all been proposed.

EIAs have been criticized for excessively limiting their scope in space and time. No accepted pro-cedure exists for determining such boundaries. The boundary refers to 'the spatial and temporal boundary of the proposal's effects'. This boundary is determined by the applicant and the lead assessor, but in practice, almost all EIAs address only direct and immediate on-site effects.

Development causes both direct and indirect effects. Consumption of goods and services, production, use and disposal of building materials and machinery, additional land use for activities of manufacturing and services, mining and refining, etc., all have environmental impacts. The indirect effects of development can be much higher than the direct effects examined by an EIA. Proposals such as airports or shipyards cause wide-ranging national and international effects, which should be covered in EIAs.

Broadening the scope of EIA can benefit the conservation of threatened species. Instead of concentrating on the project site, some EIAs employed a habitat-based approach that focused on much broader relationships among humans and the environment. As a result, alternatives that reduce the negative effects to the population of whole species, rather than local subpopulations, can be assessed.

Thissen and Agusdinata have argued that little attention is given to the systematic identification and assessment of uncertainties in environmental studies which is critical in situations where uncertainty cannot be easily reduced by doing more research. In line with this, Maier et al. have concluded on the need to consider uncertainty at all stages of the decision-making process. In such a way decisions can be made with confidence or known uncertainty. These proposals are justified on data that shows that environmental assessments fail to predict accurately the impacts observed. Tenney et al. and Wood et al. have reported evidence of the intrinsic uncertainty attached to EIAs predictions from a number of case studies worldwide. The gathered evidence consisted of comparisons between predictions in EIAs and the impacts measured during, or following project implementation. In explaining this trend, Tenney et al. have highlighted major causes such as project changes, modelling errors, errors in data and assumptions taken and bias introduced by people in the projects analyzed. Cardenas and Halman provide a comprehensive review on the issues of uncertainty in environmental impact assessments.

Environmental Manager

Environmental managers are involved in processes that *seek to* control some environmental entities in orientation to a plan or idea. Whether such control is possible, however, is contested. Examples for environmental managers range from corporate agents (corporate environmental managers) via managers of a nature reserve, to environmental and resource planning agents but, analytically seen, also involve indigenous environmental managers, farmers or environmental activists. In many accounts, hope is held that environmental managers implement grand plans or political programmes. For example, specific schools of thought like ecological modernisation but also widespread conceptions of environmental management and environmental activism presuppose human agents who have ideas, make plans and engage in action oriented at the plans' implementation. At the heart of the notion of environmental managers is, thus, a pragmatic and rational actor who optimises environments in orientation to some aim. Critical academics point out that the very idea that such managers exist and are imagined as capable of managing may well be flawed.

Corporate Environmental Managers

Steve Fineman studied UK managers and their "'green' selves and roles" in the last decade, suggesting that while environmental problems may be recognised by them, production is seen as legitimising pollution. Optimistic accounts see managers as stewards of environmental ethics. Literature differentiates different styles by managers to engage with the environment. Critics suggest that corporate environmental managers are systematically positioned in a contradictory situation in which they are supposed to be committed to competing normative orientations (e.g. profits versus environmental protection measures which do not pay off).

State Environmental Managers

State institutions can manage directly environments through their staff. And state institutions can use civil agents on their behalf. Examples for the latter are farmers who are to implement environmental regulation, citizens subject to e.g. recycling legislation or independent auditors who use laws as standards. Military agents can also act as environmental managers insofar as their action constitutes planned intervention in some environment (e.g. the burning of a forest, the destruction of streets or managing an open landscape for military training), trying to achieve military aims.

Scientists As Environmental Managers

A variety of scientists are involved directly in environmental management. Cases of ecologists acting as managers of ecosystems are known.

Study of Environmental Managers

In the field of environmental management, until now, little attention has been paid to the study of those agents who supposedly put into practices the prescriptions. For the realm of environmental management addressed by ecological modernisation, such agents can be termed *agents of ecological modernisation*.

The very notion that humans may be able to manage environments is criticised for being top-down, anthropocentric and short-sighted.

Studies of the work reality of practices of environmental managers are rarely classified as such. Studying the work reality of these managers indicates that it should be better conceptualised as situated action (for example by drawing on ethnomethodological studies of work practice) rather than rational action. Thus, when attending to managers' practices, the notion of environmental management can be reconceptualised as a *prescription*. In optimistic accounts of environmental management, the latter is often mistaken for a *description*.

Environmental Globalization

Environmental globalization refers to the internationally coordinated practices and regulations (often in the form of international treaties) regarding environmental protection.

The official logo of the Mount Everest Earth Day 20 International Peace Climb. Initiatives like
Earth Day promote international cooperation on pro-environmental initiatives,
or in other words - promote environmental globalization.

Definitions and Characteristics

Karl S. Zimmerer defined it as "the increased role of globally organized management institutions, knowledge systems and monitoring, and coordinated strategies aimed at resource, energy, and conservation issues." Alan Grainger in turn wrote that it can be understood as "an increasing spatial uniformity and contentedness in regular environmental management practices". Steven Yearley has referred to this concept as "globalization of environmental concern". Grainger also cited a study by Clark (2000), which he noted was an early treatment of the concept, and distinguished three aspects of environmental globalization: "global flows of energy, materials and organisms; formulation and global acceptance of ideas about global environment; and environmental governance" (a growing web of institutions concerned with global environment).

Environmental globalization is related to economic globalization, as economic development on a global scale has environmental impacts on such scale, which is of concern to numerous organizations and individuals. While economic globalization has environmental impacts, those impacts should not be confused with the concept of environmental globalization. In some regards, environmental globalization is in direct opposition to economic globalization, particularly when the latter is described as encouraging trade, and the former, as promoting pro-environment initiatives that are an impediment to trade. For that reason, an environmental activists might might be opposed to economic globalization, but advocate environmental globalization.

History

Grainger has discussed that environmental globalization in the context of international agreements on pro-environmental initiatives. According to him, precursors to modern environmental globalization can be found in the colonial era scientific forestry (research into how to create and restore forests). Modern initiatives contributing to environmental globalization include the 1972 United Nations Conference on the Human Environment, came from the World Bank 1980s requirements that development projects need to protect indigenous peoples and conserve biodiversity. Other examples of such initiative include treaties such like the series of International Tropical Timber Agreement treaties (1983, 1994, 2006). Therefore, unlike other main forms of globalization economic, political and cultural which were already strong in the 19th century, environmental globalization is a more recent phenomena, one that begun in earnest only in the later half of the 20th century. Similarly, Steven Yearley states that it was around that time that the environmental movement started to organize on the international scale focus on the global dimension of the issues (the first Earth Day was celebrated on 1970).

Supporters and Opponents

According to Grainger, environmental globalization (in the form of pro-environmental international initiatives) is usually supported by various non-governmental organizations and governments of developed countries, and opposed by governments of developing countries (Group of 77), which see pro-environmental initiatives as hindering their economic development. Governmental resistance to environmental globalization takes form or policy ambiguity (exemplified by countries which sign internatonal pro-environmental treaties and pass domestic pro-environmental laws, but then proceed to not enforce them) and collective resistance in forums such as United Nations to projects that would introduce stronger regulations or new institutions policing environmental issues worldwide (such as opposition to the forest-protection agreement during the Earth Summit in 1992, which was eventually downgraded from a binding to a non-binding set of Forest Principles).

World Trade Organization has also been criticized as focused on economic globalization (liberalizing trade) over concerns of environmental protection, which are seen as impeding the trade. Steven Yearley states that WTO should not be described as "anti-environmental", but its decisions have major impact on environment worldwide, and they are based primary on economic concerns, with environmental concerns being given some weight but clearly, secondary.

References

- *Rushefsky, Mark E. (2002). Public Policy in the United States at the Dawn of the Twenty-first Century (3rd ed.). New York: M.E. Sharpe, Inc. pp. 253–254. ISBN 978-0-7656-1663-0.*

- Hardman Reis, T., *Compensation for Environmental Damages Under International Law*, Kluwer Law International, The Hague, 2011, ISBN 978-90-411-3437-0.

- *Miller, Jr., G. Tyler (2003). Environmental Science: Working With the Earth (9th ed.). Pacific Grove, California: Brooks/Cole. p. G5. ISBN 0-534-42039-7.*

- *Greg Barton (2002). Empire Forestry and the Origins of Environmentalism. Cambridge University Press. p. 48. ISBN 9781139434607.*

- *Jan Marsh (1982). Back to the Land: The Pastoral Impulse in England, 1880-1914. Quartet Books. ISBN 9780704322769.*

- *Chapman, Roger (2010). Culture wars: an encyclopedia of issues, viewpoints, and voices. M.E. Sharpe, Inc. p. 162. ISBN 0-7656-1761-7.*

- *Wu and Wen (2015). Nongovernmental organizations and environmental protests: Impacts in East Asia (chapter 7 of Routledge Handbook of Environment and Society in Asia). London: Routledge. pp. 105–119. ISBN 978-0-415-65985-7.*

- *Eccleston, Charles; Doub, J. Peyton (2012). Preparing NEPA Environmental Assessments: A User's Guide to Best Professional Practices. CRC Press. ISBN 9781439808825.*

- *Karl S. Zimmerer (2006). Globalization & New Geographies of Conservation. University of Chicago Press. p. 1. ISBN 978-0-226-98344-8.*

- *Grainger, Alan (2012-01-01). Environmental Globalization. John Wiley & Sons, Ltd. doi:10.1002/9780470670590.wbeog170/full. ISBN 9780470670590.*

- *John Benyon; David Dunkerley (1 May 2014). Globalization: The Reader. Routledge. p. 54. ISBN 978-1-136-78240-4.*

- *Steve Yearly (15 April 2008). "Globalization and the Environment". In George Ritzer. The Blackwell Companion to Globalization. John Wiley & Sons. p. 240. ISBN 978-0-470-76642-2.*

- *Betty Dobratz; Lisa K Waldner; Timothy Buzzell (14 October 2015). Power, Politics, and Society: An Introduction to Political Sociology. Routledge. p. 346. ISBN 978-1-317-34529-9.*

- *Alan Grainger (31 October 2013). "Environmental Globalization and Tropical Forests". In Jan Oosthoek; Barry K. Gills. The Globalization of Environmental Crisis. Routledge. p. 63. ISBN 978-1-317-96896-2.*

5

Selected Topics of Environmental Protection

Environmental protection encompasses many different problems and is caused by various natural and artificial issues. Some selected topics of utmost significance have been presented within this chapter. It will discuss how resource depletion and ecological modernization are a part of environmental degradation. The chapter will also elaborate topics like environmentalism, environmental ethics and indigenous rights, to provide more in-depth information to the students.

Environmentalism

Environmentalism or environmental rights is a broad philosophy, ideology, and social movement regarding concerns for environmental protection and improvement of the health of the environment, particularly as the measure for this health seeks to incorporate the concerns of non-human elements. Environmentalism advocates the lawful preservation, restoration and/or improvement of the natural environment, and may be referred to as a movement to control pollution or protect plant and animal diversity. For this reason, concepts such as a land ethic, environmental ethics, biodiversity, ecology, and the biophilia hypothesis figure predominantly.

At its crux, environmentalism is an attempt to balance relations between humans and the various natural systems on which they depend in such a way that all the components are accorded a proper degree of sustainability. The exact measures and outcomes of this balance is controversial and there are many different ways for environmental concerns to be expressed in practice. Environmentalism and environmental concerns are often represented by the color green, but this association has been appropriated by the marketing industries for the tactic known as greenwashing. Environmentalism is opposed by anti-environmentalism, which says that the Earth is less fragile than some environmentalists maintain, and portrays environmentalism as overreacting to the human contribution to climate change or opposing human advancement.

Definitions

Environmentalism denotes a social movement that seeks to influence the political process by lobbying, activism, and education in order to protect natural resources and ecosystems. The word was first coined in 1922.

An *environmentalist* is a person who may speak out about our natural environment and the sustainable management of its resources through changes in public policy or individual behavior. This may include supporting practices such as informed consumption, conservation initiatives, investment in renewable resources, improved efficiencies in the materials economy, transitioning to new accounting paradigms such as Ecological economics and renewing and revitalizing our connections with non-human life.

In various ways (for example, grassroots activism and protests), environmentalists and environmental organizations seek to give the natural world a stronger voice in human affairs.

In general terms, environmentalists advocate the sustainable management of resources, and the protection (and restoration, when necessary) of the natural environment through changes in public policy and individual behavior. In its recognition of humanity as a participant in ecosystems, the movement is centered around ecology, health, and human rights.

While the term *environmentalism* focuses more on the environmental and nature-related aspects of green ideology and politics, *ecologism* as a term combines the ideology of social ecology and environmentalism. *Ecologism* as a term is more commonly used in continental European languages while *environmentalism* is more commonly used in English but the words have slightly different connotations.

History

A concern for environmental protection has recurred in diverse forms, in different parts of the world, throughout history. For example, in Europe, King Edward I of England banned the burning of sea-coal by proclamation in London in 1272, after its smoke had become a problem. The fuel was so common in England that this earliest of names for it was acquired because it could be carted away from some shores by the wheelbarrow.

Earlier in the Middle East, the Caliph Abu Bakr in the 630s commanded his army to "Bing no harm to the trees, nor burn them with fire," and "Slay not any of the enemy's flock, save for your food." Arabic medical treatises during the 9th to 13th centuries dealing with environmentalism and environmental science, including pollution, were written by Al-Kindi, Qusta ibn Luqa, Al-Razi, Ibn Al-Jazzar, al-Tamimi, al-Masihi, Avicenna, Ali ibn Ridwan, Ibn Jumay, Isaac Israeli ben Solomon, Abd-el-latif, Ibn al-Quff, and Ibn al-Nafis. Their works covered a number of subjects related to pollution, such as air pollution, water pollution, soil contamination, municipal solid waste mishandling, and environmental impact assessments of certain localities.

Early Environmental Legislation

Levels of air pollution rose during the Industrial Revolution, sparking the first modern environmental laws to be passed in the mid-19th century.

The origins of the environmental movement lay in the response to increasing levels of smoke pollution in the atmosphere during the Industrial Revolution. The emergence of great factories and the concomitant immense growth in coal consumption gave rise to an unprecedented level of air pollution in industrial centers; after 1900 the large volume of industrial chemical discharges added to the growing load of untreated human waste. The first large-scale, modern environmental laws came in the form of Britain's Alkali Acts, passed in 1863, to regulate the deleterious air pollution (gaseous hydrochloric acid) given off by the Leblanc process, used to produce soda ash. An Alkali inspector and four sub-inspectors were appointed to curb this pollution. The responsibilities of the inspectorate were gradually expanded, culminating in the Alkali Order 1958 which placed all major heavy industries that emitted smoke, grit, dust and fumes under supervision.

In industrial cities local experts and reformers, especially after 1890, took the lead in identifying environmental degradation and pollution, and initiating grass-roots movements to demand and achieve reforms. Typically the highest priority went to water and air pollution. The Coal Smoke Abatement Society was formed in 1898 making it one of the oldest environmental NGOs. It was founded by artist Sir William Blake Richmond, frustrated with the pall cast by coal smoke. Although there were earlier pieces of legislation, the Public Health Act 1875 required all furnaces and fireplaces to consume their own smoke. It also provided for sanctions against factories that emitted large amounts of black smoke. The provisions of this law were extended in 1926 with the Smoke Abatement Act to include other emissions, such as soot, ash and gritty particles and to empower local authorities to impose their own regulations.

It was, however, only under the impetus of the Great Smog of 1952 in London, which almost brought the city to a standstill and may have caused upward of 6,000 deaths that the Clean Air Act 1956 was passed and pollution in the city was finally brought to an end. Financial incentives were offered to householders to replace open coal fires with alternatives (such as installing gas fires), or for those who preferred, to burn coke instead (a byproduct of town gas production) which produces minimal smoke. 'Smoke control areas' were introduced in some towns and cities in which only smokeless fuels could be burnt and power stations were relocated away from cities. The act formed an important impetus to modern environmentalism, and caused a rethinking of the dangers of environmental degradation to people's quality of life.

The late 19th century also saw the passage of the first wildlife conservation laws. The zoologist Alfred Newton published a series of investigations into the *Desirability of establishing a 'Close-time' for the preservation of indigenous animals* between 1872 and 1903. His advocacy for legislation to protect animals from hunting during the mating season led to the formation of the Royal Society for the Protection of Birds and influenced the passage of the Sea Birds Preservation Act in 1869 as the first nature protection law in the world.

First Environmental Movements

Early interest in the environment was a feature of the Romantic movement in the early 19th century. The poet William Wordsworth travelled extensively in the Lake District and wrote that it is a "sort of national property in which every man has a right and interest who has an eye to perceive and a heart to enjoy".

John Ruskin an influential thinker who articulated the Romantic ideal of environmental protection and conservation.

Systematic efforts on behalf of the environment only began in the late 19th century; it grew out of the amenity movement in Britain in the 1870s, which was a reaction to industrialization, the growth of cities, and worsening air and water pollution. Starting with the formation of the Commons Preservation Society in 1865, the movement championed rural preservation against the encroachments of industrialisation. Robert Hunter, solicitor for the society, worked with Hardwicke Rawnsley, Octavia Hill, and John Ruskin to lead a successful campaign to prevent the construction of railways to carry slate from the quarries, which would have ruined the unspoilt valleys of Newlands and Ennerdale. This success led to the formation of the Lake District Defence Society (later to become The Friends of the Lake District).

In 1893 Hill, Hunter and Rawnsley agreed to set up a national body to coordinate environmental conservation efforts across the country; the "National Trust for Places of Historic Interest or Natural Beauty" was formally inaugurated in 1894. The organisation obtained secure footing through the 1907 National Trust Bill, which gave the trust the status of a statutory corporation. and the bill was passed in August 1907.

An early "Back-to-Nature" movement, which anticipated the romantic ideal of modern environmentalism, was advocated by intellectuals such as John Ruskin, William Morris, George Bernard Shaw and Edward Carpenter, who were all against consumerism, pollution and other activities that were harmful to the natural world. The movement was a reaction to the urban conditions of the industrial towns, where sanitation was awful, pollution levels intolerable and housing terribly cramped. Idealists championed the rural life as a mythical Utopia and advocated a return to it. John Ruskin argued that people should return to a *small piece of English ground, beautiful, peaceful, and fruitful. We will have no steam engines upon it . . . we will have plenty of flowers and vegetables . . . we will have some music and poetry; the children will learn to dance to it and sing it.*

Practical ventures in the establishment of small cooperative farms were even attempted and old rural traditions, without the "taint of manufacture or the canker of artificiality", were enthusiastically revived, including the Morris dance and the maypole.

These ideas also inspired various environmental groups in the UK, such as the Royal Society for the Protection of Birds, established in 1889 by Emily Williamson as a protest group to campaign for greater protection for the indigenous birds of the island. The Society attracted growing support from the suburban middle-classes as well as support from many other influential figures, such as the ornithologist Professor Alfred Newton. By 1900, public support for the organisation had grown, and it had over 25,000 members. The Garden city movement incorporated many environmental concerns into its urban planning manifesto; the Socialist League and The Clarion movement also began to advocate measures of nature conservation.

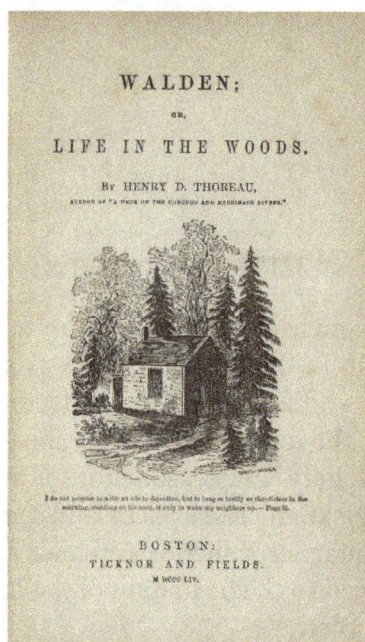

Original title page of *Walden* by Henry David Thoreau.

The movement in the United States began in the late 19th century, out of concerns for protecting the natural resources of the West, with individuals such as John Muir and Henry David Thoreau making key philosophical contributions. Thoreau was interested in peoples' relationship with nature and studied this by living close to nature in a simple life. He published his experiences in the book *Walden*, which argues that people should become intimately close with nature. Muir came to believe in nature's inherent right, especially after spending time hiking in Yosemite Valley and studying both the ecology and geology. He successfully lobbied congress to form Yosemite National Park and went on to set up the Sierra Club in 1892. The conservationist principles as well as the belief in an inherent right of nature were to become the bedrock of modern environmentalism.

In the 20th century, environmental ideas continued to grow in popularity and recognition. Efforts were starting to be made to save some wildlife, particularly the American bison. The death of the last passenger pigeon as well as the endangerment of the American bison helped to focus the minds of conservationists and popularize their concerns. In 1916 the National Park Service was founded by US President Woodrow Wilson.

The Forestry Commission was set up in 1919 in Britain to increase the amount of woodland in Britain by buying land for afforestation and reforestation. The commission was also tasked with promoting forestry and the production of timber for trade. During the 1920s the Commission focused on acquiring land to begin planting out new forests; much of the land was previously used for agricultural purposes. By 1939 the Forestry Commission was the largest landowner in Britain.

During the 1930s the Nazis had elements that were supportive of animal rights, zoos and wildlife, and took several measures to ensure their protection. In 1933 the government created a stringent animal-protection law and in 1934, *Das Reichsjagdgesetz* (The Reich Hunting Law) was enacted which limited hunting. Several Nazis were environmentalists (notably Rudolf Hess), and species protection and animal welfare were significant issues in the regime. In 1935, the regime enacted the "Reich Nature Protection Act" (*Reichsnaturschutzgesetz*). The concept of the *Dauerwald* (best translated as the "perpetual forest") which included concepts such as forest management and protection was promoted and efforts were also made to curb air pollution.

In 1949, *A Sand County Almanac* by Aldo Leopold was published. It explained Leopold's belief that humankind should have moral respect for the environment and that it is unethical to harm it. The book is sometimes called the most influential book on conservation.

Throughout the 1950s, 1960s, 1970s and beyond, photography was used to enhance public awareness of the need for protecting land and recruiting members to environmental organizations. David Brower, Ansel Adams and Nancy Newhall created the Sierra Club Exhibit Format Series, which helped raise public environmental awareness and brought a rapidly increasing flood of new members to the Sierra Club and to the environmental movement in general. "This Is Dinosaur" edited by Wallace Stegner with photographs by Martin Litton and Philip Hyde prevented the building of dams within Dinosaur National Monument by becoming part of a new kind of activism called environmentalism that combined the conservationist ideals of Thoreau, Leopold and Muir with hard-hitting advertising, lobbying, book distribution, letter writing campaigns, and more. The powerful use of photography in addition to the written word for conservation dated back to the creation of Yosemite National Park, when photographs persuaded Abraham Lincoln to preserve the beautiful glacier carved landscape for all time. The Sierra Club Exhibit Format Series galvanized public opposition to building dams in the Grand Canyon and protected many other national treasures. The Sierra Club often led a coalition of many environmental groups including the Wilderness Society and many others. After a focus on preserving wilderness in the 1950s and 1960s, the Sierra Club and other groups broadened their focus to include such issues as air and water pollution, population concern, and curbing the exploitation of natural resources.

Post-War Expansion

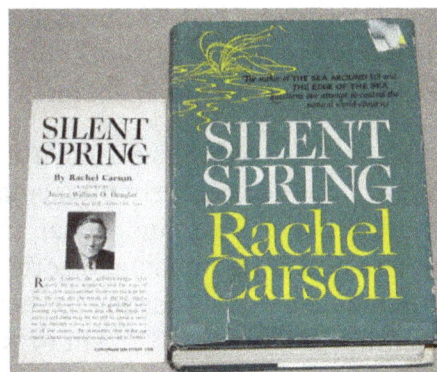

Silent Spring by Rachael Carson, published in 1962, included an endorsement by William O. Douglas.

In 1962, *Silent Spring* by American biologist Rachel Carson was published. The book cataloged the environmental impacts of the indiscriminate spraying of DDT in the US and questioned the logic of releasing large amounts of chemicals into the environment without fully understanding their effects on ecology or human health. The book suggested that DDT and other pesticides may cause cancer and that their agricultural use was a threat to wildlife, particularly birds. The resulting public concern led to the creation of the United States Environmental Protection Agency in 1970 which subsequently banned the agricultural use of DDT in the US in 1972. The limited use of DDT in disease vector control continues to this day in certain parts of the world and remains controversial. The book's legacy was to produce a far greater awareness of environmental issues and interest into how people affect the environment. With this new interest in environment came interest in problems such as air pollution and petroleum spills, and environmental interest grew. New pressure groups formed, notably Greenpeace and Friends of the Earth (US), as well as notable local organizations such as the Wyoming Outdoor Council, which was founded in 1967.

In the 1970s, the environmental movement gained rapid speed around the world as a productive outgrowth of the counterculture movement.

The world's first political parties to campaign on a predominantly environmental platform were the United Tasmania Group Tasmania, Australia and the Values Party of New Zealand. The first green party in Europe was the Popular Movement for the Environment, founded in 1972 in the Swiss canton of Neuchâtel. The first national green party in Europe was PEOPLE, founded in Britain in February 1973, which eventually turned into the Ecology Party, and then the Green Party.

Protection of the environment also became important in the developing world; the Chipko movement was formed in India under the influence of Mohandas Gandhi and they set up peaceful resistance to deforestation by literally hugging trees (leading to the term "tree huggers"). Their peaceful methods of protest and slogan "ecology is permanent economy" were very influential.

Another milestone in the movement was the creation of an Earth Day. Earth Day was first observed in San Francisco and other cities on March 21, 1970, the first day of spring. It was created to give awareness to environmental issues. On March 21, 1971, United Nations Secretary-General U Thant spoke of a spaceship Earth on Earth Day, hereby referring to the ecosystem services the earth supplies to us, and hence our obligation to protect it (and with it, ourselves). Earth Day is now coordinated globally by the Earth Day Network, and is celebrated in more than 175 countries every year.

The UN's first major conference on international environmental issues, the United Nations Conference on the Human Environment (also known as the Stockholm Conference), was held on June 5–16, 1972. It marked a turning point in the development of international environmental politics.

By the mid-1970s, many felt that people were on the edge of environmental catastrophe. The Back-to-the-land movement started to form and ideas of environmental ethics joined with anti-Vietnam War sentiments and other political issues. These individuals lived outside normal society and started to take on some of the more radical environmental theories such as deep ecology. Around this time more mainstream environmentalism was starting to show force with the signing of the Endangered Species Act in 1973 and the formation of CITES in 1975. Significant amendments were also enacted to the United States Clean Air Act and Clean Water Act.

In 1979, James Lovelock, a British scientist, published *Gaia: A new look at life on Earth*, which put forth the Gaia hypothesis; it proposes that life on earth can be understood as a single organism. This became an important part of the Deep Green ideology. Throughout the rest of the history of environmentalism there has been debate and argument between more radical followers of this Deep Green ideology and more mainstream environmentalists.

Today

Environmentalism continues to evolve to face up to new issues such as global warming, overpopulation and genetic engineering.

Recent research demonstrates a precipitous decline in the public's interest in 19 different areas of environmental concern. Americans are less likely be actively participating in an environmental movement or organization and more likely to identify as "unsympathetic" to an environmental movement then in 2000. This is likely a lingering factor of the Great Recession in 2008. Since 2005 the percentage of Americans agreeing that the environment should be given priority over economic growth has dropped 10 points, in contrast, those feeling that growth should be given priority "even if the environment suffers to some extent" has risen 12 percent. These numbers point to the growing complexity of environmentalism and its relationship to economics.

Environmental Movement

Before flue-gas desulfurization was installed, the air-polluting emissions from this power plant in New Mexico contained excessive amounts of sulfur dioxide.

The *environmental movement* (a term that sometimes includes the conservation and green movements) is a diverse scientific, social, and political movement. Though the movement is represented by a range of organizations, because of the inclusion of environmentalism in the classroom curriculum, the environmental movement has a younger demographic than is common in other social movements (see green seniors).

Environmentalism as a movement covers broad areas of institutional oppression, including for example: consumption of ecosystems and natural resources into waste, dumping waste into disadvantaged communities, air pollution, water pollution, weak infrastructure, exposure of organic life to toxins, mono-culture, anti-polythene drive (jhola movement) and various other focuses. Because of these divisions, the environmental movement can be categorized into these primary focuses: environmental science, environmental activism, environmental advocacy, and environmental justice.

Free Market Environmentalism

Free market environmentalism is a theory that argues that the free market, property rights, and tort law provide the best tools to preserve the health and sustainability of the environment. It considers environmental stewardship to be natural, as well as the expulsion of polluters and other aggressors through individual and class action. It has been supported by libertarians and many conservatives.

Evangelical Environmentalism

Evangelical environmentalism is an environmental movement in the United States in which some Evangelicals have emphasized biblical mandates concerning humanity's role as steward and subsequent responsibility for the caretaking of Creation. While the movement has focused on different environmental issues, it is best known for its focus of addressing climate action from a biblically grounded theological perspective. The Evangelical Climate Initiative argues that human-induced climate change will have severe consequences and impact the poor the hardest, and that God's mandate to Adam to care for the Garden of Eden also applies to evangelicals today, and that it is therefore a moral obligation to work to mitigate climate impacts and support communities in adapting to change.

Preservation and Conservation

Environmental preservation in the United States and other parts of the world, including Australia, is viewed as the setting aside of natural resources to prevent damage caused by contact with humans or by certain human activities, such as logging, mining, hunting, and fishing, often to replace them with new human activities such as tourism and recreation. Regulations and laws may be enacted for the preservation of natural resources.

Organizations and Conferences

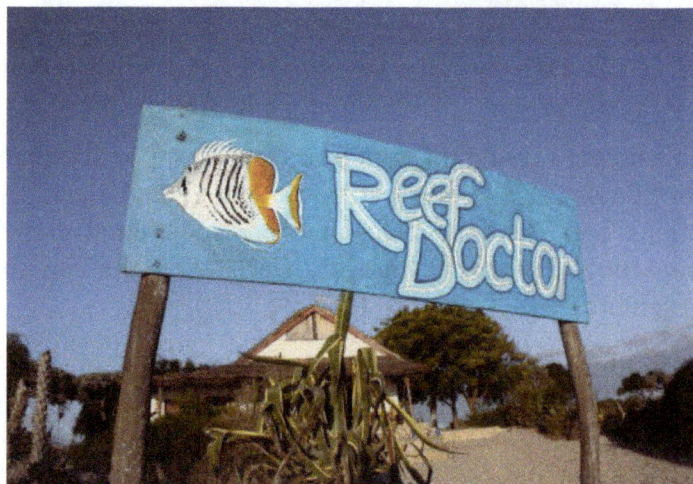

Reef doctor work station in Ifaty, Madagascar

Environmental organizations can be global, regional, national or local; they can be government-run or private (NGO). Environmentalist activity exists in almost every country. Moreover, groups dedicated to community development and social justice also focus on environmental concerns.

There are some volunteer organizations. For example, Ecoworld and Paryawaran Sachetak Samiti are environmental organizations which are based on teamwork and volunteer work. Some US environmental organizations, among them the Natural Resources Defense Council and the Environmental Defense Fund, specialize in bringing lawsuits (a tactic seen as particularly useful in that country). Other groups, such as the US-based National Wildlife Federation, the Nature Conservancy, and The Wilderness Society, and global groups like the World Wide Fund for Nature and Friends of the Earth, disseminate information, participate in public hearings, lobby, stage demonstrations, and may purchase land for preservation. Statewide nonprofit organizations such as the Wyoming Outdoor Council often collaborate with these national organizations and employ similar strategies. Smaller groups, including Wildlife Conservation International, conduct research on endangered species and ecosystems. More radical organizations, such as Greenpeace, Earth First!, and the Earth Liberation Front, have more directly opposed actions they regard as environmentally harmful. While Greenpeace is devoted to nonviolent confrontation as a means of bearing witness to environmental wrongs and bringing issues into the public realm for debate, the underground *Earth Liberation Front* engages in the clandestine destruction of property, the release of caged or penned animals, and other criminal acts. Such tactics are regarded as unusual within the movement, however.

On an international level, concern for the environment was the subject of a United Nations Conference on the Human Environment in Stockholm in 1972, attended by 114 nations. Out of this meeting developed UNEP (United Nations Environment Programme) and the follow-up United Nations Conference on Environment and Development in 1992. Other international organizations in support of environmental policies development include the Commission for Environmental Cooperation (as part of NAFTA), the European Environment Agency (EEA), and the Intergovernmental Panel on Climate Change (IPCC).

Environmental Protests

Climate activists blockade British Airports Authority's headquarters for day of action

"March Against Monsanto", Vancouver, Canada, May 25, 2013

Notable environmental protests and campaigns include:

- 2010 Xinfa aluminum plant protest
- Anti-WAAhnsinns Festival
- Camp for Climate Action
- Campaign against Climate Change
- Climate Rush
- Cofán people oil drilling protest (Ecuador)
- Earth Day
- Earth First!
- Earthlife Africa
- Global Day of Action
- Gurindji Strike
- Hands off our Forest
- Homes before Roads
- Kupa Piti Kungka Tjuta
- Love Canal protests
- March Against Monsanto
- Nevada Desert Experience
- Plane Mad

- Plane Stupid
- Qidong protest
- Save Manapouri Campaign
- Say Yes demonstrations
- Shifang protest
- Stop Climate Chaos

Environmentalists

Notable advocates for environmental protection and sustainability include:

- Edward Abbey (author)
- David Attenborough (broadcaster, naturalist)
- John James Audubon (naturalist)
- David Bellamy (botanist)
- Wendell Berry (farmer, philosopher)
- David Brower (writer, activist)
- Lester Brown (environmental analyst, author)
- Kevin Buzzacott (Aboriginal activist)
- Helen Caldicott (medical doctor)
- Rachel Carson (biologist, writer)
- Prince Charles (British Royal Family member)
- Barry Commoner (biologist, politician)
- Jacques-Yves Cousteau (explorer, ecologist)
- Peter Dauvergne (political scientist)
- Leonardo Dicaprio (actor and environmentalist)
- Paul R. Ehrlich (population biologist)
- Hans-Josef Fell (German Green Party member)
- Jane Fonda (actor)
- Mizuho Fukushima (politician, activist)
- Peter Garrett (musician, politician)
- Al Gore (former Vice President of the United States)
- James Hansen (scientist)

- Garrett Hardin (ecologist, ecophilosopher)
- Denis Hayes (environmentalist and solar power advocate)
- Tetsunari Iida (sustainable energy advocate)
- Aldo Leopold (ecologist)
- A. Carl Leopold (plant physiologist)
- James Lovelock (scientist)
- Amory Lovins (energy policy analyst)
- Hunter Lovins (environmentalist)
- Caroline Lucas (politician)
- Bill McKibben (writer, activist)
- David McTaggart (activist)
- Chico Mendes (activist)
- George Monbiot (journalist)
- John Muir (naturalist, activist)
- Ralph Nader (activist)
- Gaylord Nelson (politician)
- Alan Pears (environmental consultant and energy efficiency pioneer)
- Gifford Pinchot (first chief of the USFS)
- Jonathon Porritt (politician)
- John Wesley Powell (second director of the USGS)
- Barbara Pyle (documentarian and executive producer of *Captain Planet and the Planeteers*)
- Phil Radford (environmental, clean energy and democracy advocate, Greenpeace Executive Director)
- Bonnie Raitt (musician)
- Theodore Roosevelt (former President of the United States)
- E. F. Schumacher (author of *Small is Beautiful*)
- Swami Sundaranand {Yogi, photographer, and mountaineer}
- Cass Sunstein (environmental lawyer)
- David Suzuki (scientist, broadcaster)
- Henry David Thoreau (writer, philosopher)

- Stewart Udall (former United States Secretary of the Interior)

- Jo Valentine (politician and activist)

- Dominique Voynet (politician and environmentalist)

- Gabriel Willow (environmental educator, naturalist)

- Howard Zahniser (author of the 1964 Wilderness Act)

Usage in Popular Culture

- The popular media have been used to convey conservation messages in the U.S. For instance, the U.S. Forest Service created Smokey the Bear in 1944; he appeared in countless posters, radio and television programs, movies, press releases, and other guises to warn about forest fires. The comic strip Mark Trail, by environmentalist Ed Dodd, began in 1946; it still appears weekly in 175 newspapers. Another example is the children's animated show *Captain Planet and the Planeteers*, created by Ted Turner and Barbara Pyle in 1989 to inform kids on environmental issues. The show aired for six seasons and 113 episodes, in 100 countries worldwide from 1990 to 1996.

- Miss Earth is one of the three largest international beauty pageants alongside Miss Universe and Miss World that promotes Environmental Awareness. The reigning titleholders dedicate their year to promote specific projects and often address issues concerning the environment and other global issues through school tours, tree planting activities, street campaigns, coastal clean ups, speaking engagements, shopping mall tours, media guesting, environmental fair, storytelling programs, eco-fashion shows, and other environmental activities. The Miss Earth winner is the spokesperson for the Miss Earth Foundation, the United Nations Environment Programme (UNEP) and other environmental organizations. The Miss Earth Foundation also works with the environmental departments and ministries of participating countries, various private sectors and corporations, as well as Greenpeace and the World Wildlife Foundation (WWF).

An Alternative View

Many environmentalists believe that human interference with 'nature' should be restricted or minimised as a matter of urgency (for the sake of life, or the planet, or just for the benefit of the human species), whereas environmental skeptics and anti-environmentalists do not believe that there is such a need. One can also regard oneself as an environmentalist and believe that human 'interference' with 'nature' should be *increased*. Nevertheless, there is a risk that the shift from emotional environmentalism into the technical management of natural resources and hazards could decrease the touch of humans with nature, leading to less concern with environment preservation.

Resource Depletion

Resource depletion is the consumption of a resource faster than it can be replenished. Natural resources are commonly divided between renewable resources and non-renewable resources (see also mineral resource classification). Use of either of these forms of resources beyond their rate of replacement is considered to be resource depletion.

Resource depletion is most commonly used in reference to farming, fishing, mining, water usage, and consumption of fossil fuels.

Causes

- Aquifer depletion

- Habitat degradation leads to the loss of biodiversity (i.e. species and ecosystems with its ecosystem services)

- Irrigation

- Mining for fossil fuels and minerals

- Overconsumption, excessive or unnecessary use of resources

- Overpopulation

- Pollution or contamination of resources

- Slash-and-burn agricultural practices, currently occurring in many developing countries

- Soil erosion

- Technological and industrial development

- Deforestation.

Minerals

Minerals are needed to provide food, clothing, and housing. A United States Geological Survey (USGS) study found a significant long-term trend over the 20th century for non-renewable resources such as minerals to supply a greater proportion of the raw material inputs to the non-fuel, non-food sector of the economy; an example is the greater consumption of crushed stone, sand, and gravel used in construction.

Large-scale exploitation of minerals began in the Industrial Revolution around 1760 in England and has grown rapidly ever since. Technological improvements have allowed humans to dig deeper and access lower grades and different types of ore over that time. Virtually all basic industrial metals (copper, iron, bauxite, etc.), as well as rare earth minerals, face production output limitations from time to time, because supply involves large up-front investments and is therefore slow to respond to rapid increases in demand.

Minerals projected by some to enter production decline during the next 20 years:

- Gas (2023)

- Copper (2024). Data from the United States Geological Survey (USGS) suggest that it is very unlikely that copper production will peak before 2040.

- Zinc. New developments in hydrometallurgy have transformed non-sulphide zinc deposits (largely ignored until now) into large low cost reserves.

Minerals projected by some to enter production decline during the present century:

- Aluminium (2057)

- Coal (2060)

- Iron (2068).

Such projections may change, as new discoveries are made and typically misinterpret available data on Mineral Resources and Mineral Reserves.

Oil

Peak oil is the period when the maximum rate of global petroleum extraction is reached, after which the rate of production enters terminal decline. It relates to a long-term decline in the available supply of petroleum. This, combined with increasing demand, will significantly increase the worldwide prices of petroleum derived products. Most significant will be the availability and price of liquid fuel for transportation.

The United States Department of Energy in the Hirsch report indicates that "The peaking of world oil production presents the U. S. and the world with an unprecedented risk management problem. As peaking is approached, liquid fuel prices and price volatility will increase dramatically, and, without timely mitigation, the economic, social, and political costs will be unprecedented. Viable mitigation options exist on both the supply and demand sides, but to have substantial impact, they must be initiated more than a decade in advance of peaking."

Deforestation

Deforestation is the clearing of forests by logging or burning of trees and plants in a forested area. As a result of deforestation, presently about one half of the forests that once covered Earth have been destroyed. It occurs for many different reasons, and it has several negative implications on the atmosphere and the quality of the land in and surrounding the forest.

Causes

One of the main causes of deforestation is clearing forests for agricultural reasons. As the population of developing areas, especially near rainforests, increases, the need for land for farming becomes more and more important. For most people, a forest has no value when its resources aren't being used, so the incentives to deforest these areas outweigh the incentives to preserve the forests. For this reason, the economic value of the forests is very important for the developing countries.

Environmental Impact

Because deforestation is so extensive, it has made several significant impacts on the environment, including carbon dioxide in the atmosphere, changing the water cycle, an increase in soil erosion, and a decrease in biodiversity. Deforestation is often cited as a cause of global warming. Because trees and plants remove carbon dioxide and emit oxygen into the atmosphere, the reduction of forests contribute to about 12% of anthropogenic carbon dioxide emissions. One of the most pressing issues that deforestation creates is soil erosion. The removal of trees causes higher rates of erosion, increasing risks of landslides, which is a direct threat to many people living close to deforested areas. As forests get destroyed, so does the habitat for millions of animals. It is estimated that 80% of the world's known biodiversity lives in the rainforests, and the destruction of these rainforests is accelerating extinction at an alarming rate.

Controlling Deforestation

The United Nations and the World Bank created programs such as Reducing Emissions from Deforestation and Forest Degradation (REDD), which works especially with developing countries to use subsidies or other incentives to encourage citizens to use the forest in a more sustainable way. In addition to making sure that emissions from deforestation are kept to a minimum, an effort to educate people on sustainability and helping them to focus on the long-term risks is key to the success of these programs. The New York Declaration on Forests and its associated actions promotes reforestation, which is being encouraged in many countries in an attempt to repair the damage that deforestation has done.

Wetlands

Wetlands are areas that are often saturated by enough surface or groundwater to sustain vegetation that is usually adapted to saturated soil conditions, such as cattails, bulrushes, red maples, wild rice, blackberries, cranberries, and peat moss. Because some varieties of wetlands are rich in minerals and nutrients and provide many of the advantages of both land and water environments they contain diverse species and possibly even form a food chain. When human activities take away resources many species are affected. An ecosystem contains many species.

Years ago people assumed wetlands were useless so it was not a large concern when they were being dug up. Many people want to use them for developing homes etc. On the other side of the argument, people believe that the wetlands are a vital source for other life forms and a part of the life cycle.

Wetlands provide services for:

1. Food and habitat

2. Improving water quality

3. Commercial fishing

4. Floodwater reduction

5. Shoreline stabilization

6. Recreation

Some loss of wetlands resulted from natural causes such as erosion, sedimentation (the buildup of soil by the settling of fine particles over a long period of time), subsidence (the sinking of land because of diminishing underground water supplies), and a rise in the sea level.

Ecological Modernization

Ecological modernization is an optimistic school of thought in the social sciences that argues that the economy benefits from moves towards environmentalism. It has gained increasing attention among scholars and policymakers in the last several decades internationally. It is an analytical approach as well as a policy strategy and environmental discourse (Hajer, 1995).

Origins and Key Elements

Ecological modernization emerged in the early 1980s within a group of scholars at Free University and the Social Science Research Centre in Berlin, among them Joseph Huber, Martin Jänicke (de) and Udo E. Simonis. Various authors pursued similar ideas at the time, e.g. Arthur H. Rosenfeld, Amory Lovins, Donald Huisingh, René Kemp, or Ernst Ulrich von Weizsäcker. Further substantial contributions were made by Arthur P.J. Mol, Gert Spaargaren and David A Sonnenfeld (Mol and Sonnenfeld, 2000; Mol, 2001).

One basic assumption of ecological modernization relates to environmental readaptation of economic growth and industrial development. On the basis of enlightened self-interest, economy and ecology can be favourably combined: Environmental productivity, i.e. productive use of natural resources and environmental media (air, water, soil, ecosystems), can be a source of future growth and development in the same way as labour productivity and capital productivity. This includes increases in energy and resource efficiency as well as product and process innovations such as environmental management and sustainable supply chain management, clean technologies, benign substitution of hazardous substances, and product design for environment. Radical innovations in these fields can not only reduce quantities of resource turnover and emissions, but also change the quality or structure of the industrial metabolism. In the co-evolution of humans and nature, and in order to upgrade the environment's carrying capacity, ecological modernization gives humans an active role to play, which may entail conflicts with nature conservation.

There are different understandings of the scope of ecological modernization - whether it is just about techno-industrial progress and related aspects of polity and economy, and to what extent it also includes cultural aspects (ecological modernization of mind, value orientiations, attitudes, behaviour and lifestyles). Similarly, there is some pluralism as to whether ecological modernization would need to rely mainly on government, or markets and entrepreneurship, or civil society, or some sort of multi-level governance combining the three. Some scholars explicitly refer to general modernization theory as well as non-Marxist world-system theory, others don't.

Ultimately, however, there is a common understanding that ecological modernization will have to result in innovative structural change. So research is now still more focused on environmental innovations, or eco-innovations, and the interplay of various societal factors (scientific, economic, institutional, legal, political, cultural) which foster or hamper such innovations (Klemmer et al.,

1999; Huber, 2004; Weber and Hemmelskamp, 2005; Olsthoorn and Wieczorek, 2006).

Ecological modernization shares a number of features with neighbouring, overlapping approaches. Among the most important are

- the concept of sustainable development

- the approach of industrial metabolism (Ayres and Simonis, 1994)

- the concept of industrial ecology (Socolow, 1994).

Additional Elements

A special topic of ecological modernization research during recent years was *sustainable household*, i.e. environment-oriented reshaping of lifestyles, consumption patterns, and demand-pull control of supply chains (Vergragt, 2000; OECD 2002). Some scholars of ecological modernization share an interest in industrial symbiosis, i.e. inter-site recycling that helps to reduce the consumption of resources via increasing efficiency (i.e. pollution prevention, waste reduction), typically by taking externalities from one economic production process and using them as raw material inputs for another (Christoff, 1996). Ecological modernization also relies on product life-cycle assessment and the analysis of materials and energy flows. In this context, ecological modernization promotes 'cradle to cradle' manufacturing (Braungart and McDonough, 2002), contrasted against the usual 'cradle to grave' forms of manufacturing - where waste is not re-integrated back into the production process. Another special interest in the ecological modernization literature has been the role of social movements and the emergence of civil society as a key agent of change (Fisher and Freudenburg, 2001).

As a strategy of change, some forms of ecological modernization may be favored by business interests because they seemingly meet the triple bottom line of economics, society, and environment, which, it is held, underpin sustainability, yet do not challenge free market principles. This contrasts with many environmental movement perspectives, which regard free trade and its notion of business self-regulation as part of the problem, or even an origin of environmental degradation. Under ecological modernization, the state is seen in a variety of roles and capacities: as the enabler for markets that help produce the technological advances via competition; as the regulatory (see regulation) medium through which corporations are forced to 'take back' their various wastes and re-integrate them in some manner into the production of new goods and services (e.g. the way that car corporations in Germany are required to accept back cars they manufactured once those vehicles have reached the end of their product lifespan); and in some cases as an institution that is incapable of addressing critical local, national, and global environmental problems. In the latter case, ecological modernization shares with Ulrich Beck (1999, 37-40) and others notions of the necessity of emergence of new forms of environmental governance, sometimes referred to as subpolitics or political modernization, where the environmental movement, community groups, businesses, and other stakeholders increasingly take on direct and leadership roles in stimulating environmental transformation. Political modernization of this sort requires certain supporting norms and institutions such as a free, independent, or at least critical press, basic human rights of expression, organization, and assembly, etc. New media such as the Internet greatly facilitate this.

Criticisms

Critics argue that ecological modernization will fail to protect the environment and does nothing to alter the impulses within the capitalist economic mode of production (see capitalism) that inevitably lead to environmental degradation (Foster, 2002). As such, it is just a form of 'green-washing'. Critics question whether technological advances alone can achieve resource conservation and better environmental protection, particularly if left to business self-regulation practices (York and Rosa, 2003). For instance, many technological improvements are currently feasible but not widely utilized. The most environmentally friendly product or manufacturing process (which is often also the most economically efficient) is not always the one automatically chosen by self-regulating corporations (e.g. hydrogen or biofuel vs. peak oil). In addition, some critics have argued that ecological modernization does not redress gross injustices that are produced within the capitalist system, such as environmental racism - where people of color and low income earners bear a disproportionate burden of environmental harm such as pollution, and lack access to environmental benefits such as parks, and social justice issues such as eliminating unemployment (Bullard, 1993; Gleeson and Low, 1999; Harvey, 1996) - environmental racism is also referred to as issues of the asymmetric distribution of environmental resources and services (Everett & Neu, 2000). Moreover, the theory seems to have limited global efficacy, applying primarily to its countries of origin - Germany and the Netherlands, and having little to say about the developing world (Fisher and Freudenburg, 2001). Perhaps the harshest criticism though, is that ecological modernization is predicated upon the notion of 'sustainable growth', and in reality this is not possible because growth entails the consumption of natural and human capital at great costs to ecosystems and societies.

Ecological modernization, its effectiveness and applicability, strengths and limitations, remains a dynamic and contentious area of environmental social science research and policy discourse in the early 21st century.

Environmental Ethics

Environmental ethics is the part of environmental philosophy which considers extending the traditional boundaries of ethics from solely including humans to including the non-human world. It exerts influence on a large range of disciplines including environmental law, environmental sociology, ecotheology, ecological economics, ecology and environmental geography.

There are many ethical decisions that human beings make with respect to the environment. For example:

- Should humans continue to clear cut forests for the sake of human consumption?

- Why should humans continue to propagate its species, and life itself?

- Should humans continue to make gasoline powered vehicles?

- What environmental obligations do humans need to keep for future generations?

- Is it right for humans to knowingly cause the extinction of a species for the convenience of humanity?

- How should humans best use and conserve the space environment to secure and expand life?

The academic field of environmental ethics grew up in response to the work of scientists such as Rachel Carson and events such as the first Earth Day in 1970, when environmentalists started urging philosophers to consider the philosophical aspects of environmental problems. Two papers published in *Science* had a crucial impact: Lynn White's "The Historical Roots of our Ecologic Crisis" (March 1967) and Garrett Hardin's "The Tragedy of the Commons" (December 1968). Also influential was Garett Hardin's later essay called "Exploring New Ethics for Survival", as well as an essay by Aldo Leopold in his *A Sand County Almanac*, called "The Land Ethic," in which Leopold explicitly claimed that the roots of the ecological crisis were philosophical (1949).

The first international academic journals in this field emerged from North America in the late 1970s and early 1980s – the US-based journal *Environmental Ethics* in 1979 and the Canadian-based journal *The Trumpeter: Journal of Ecosophy* in 1983. The first British based journal of this kind, *Environmental Values*, was launched in 1992.

Marshall's Categories of Environmental Ethics

Some scholars have tried to categorise the various ways the natural environment is valued. Alan Marshall and Michael Smith are two examples of this, as cited by Peter Vardy in "The Puzzle of Ethics". According to Marshall, three general ethical approaches have emerged over the last 40 years: Libertarian Extension, the Ecologic Extension and Conservation Ethics.

Libertarian Extension

Marshall's Libertarian extension echoes a civil liberty approach (i.e. a commitment to extend equal rights to all members of a community). In environmentalism, though, the community is generally thought to consist of non-humans as well as humans.

Andrew Brennan was an advocate of ecologic humanism (eco-humanism), the argument that all ontological entities, animate and in-animate, can be given ethical worth purely on the basis that they exist. The work of Arne Næss and his collaborator Sessions also falls under the libertarian extension, although they preferred the term "deep ecology". Deep ecology is the argument for the intrinsic value or inherent worth of the environment – the view that it is valuable in itself. Their argument, incidentally, falls under both the libertarian extension and the ecologic extension.

Peter Singer's work can be categorized under Marshall's 'libertarian extension'. He reasoned that the "expanding circle of moral worth" should be redrawn to include the rights of non-human animals, and to not do so would be guilty of speciesism. Singer found it difficult to accept the argument from intrinsic worth of a-biotic or "non-sentient" (non-conscious) entities, and concluded in his first edition of "Practical Ethics" that they should not be included in the expanding circle of moral worth. This approach is essentially then, bio-centric. However, in a later edition of "Practical Ethics" after the work of Næss and Sessions, Singer admits that, although unconvinced by deep ecology, the argument from intrinsic value of non-sentient entities is plausible, but at best problematic. Singer advocated a humanist ethics.

Ecologic Extension

Alan Marshall's category of ecologic extension places emphasis not on human rights but on the recognition of the fundamental interdependence of all biological (and some abiological) entities and their essential diversity. Whereas Libertarian Extension can be thought of as flowing from a political reflection of the natural world, Ecologic Extension is best thought of as a scientific reflection of the natural world. Ecological Extension is roughly the same classification of Smith's eco-holism, and it argues for the intrinsic value inherent in collective ecological entities like ecosystems or the global environment as a whole entity. Holmes Rolston, among others, has taken this approach.

This category might include James Lovelock's Gaia hypothesis; the theory that the planet earth alters its geo-physiological structure over time in order to ensure the continuation of an equilibrium of evolving organic and inorganic matter. The planet is characterized as a unified, holistic entity with ethical worth of which the human race is of no particular significance in the long run.

Conservation Ethics

Marshall's category of 'conservation ethics' is an extension of use-value into the non-human biological world. It focuses only on the worth of the environment in terms of its utility or usefulness to humans. It contrasts the intrinsic value ideas of 'deep ecology', hence is often referred to as 'shallow ecology', and generally argues for the preservation of the environment on the basis that it has extrinsic value – instrumental to the welfare of human beings. Conservation is therefore a means to an end and purely concerned with mankind and inter-generational considerations. It could be argued that it is this ethic that formed the underlying arguments proposed by Governments at the Kyoto summit in 1997 and three agreements reached in Rio in 1992.

Humanist Theories

Following the bio-centric and eco-holist theory distinctions, Michael Smith further classifies Humanist theories as those that require a set of criteria for moral status and ethical worth, such as sentience. This applies to the work of Peter Singer who advocated a hierarchy of value similar to the one devised by Aristotle which relies on the ability to reason. This was Singer's solution to the problem that arises when attempting to determine the interests of a non-sentient entity such as a garden weed.

Singer also advocated the preservation of "world heritage sites," unspoilt parts of the world that acquire a "scarcity value" as they diminish over time. Their preservation is a bequest for future generations as they have been inherited from human's ancestors and should be passed down to future generations so they can have the opportunity to decide whether to enjoy unspoilt countryside or an entirely urban landscape. A good example of a world heritage site would be the tropical rainforest, a very specialist ecosystem that has taken centuries to evolve. Clearing the rainforest for farmland often fails due to soil conditions, and once disturbed, can take thousands of years to regenerate.

Applied Theology

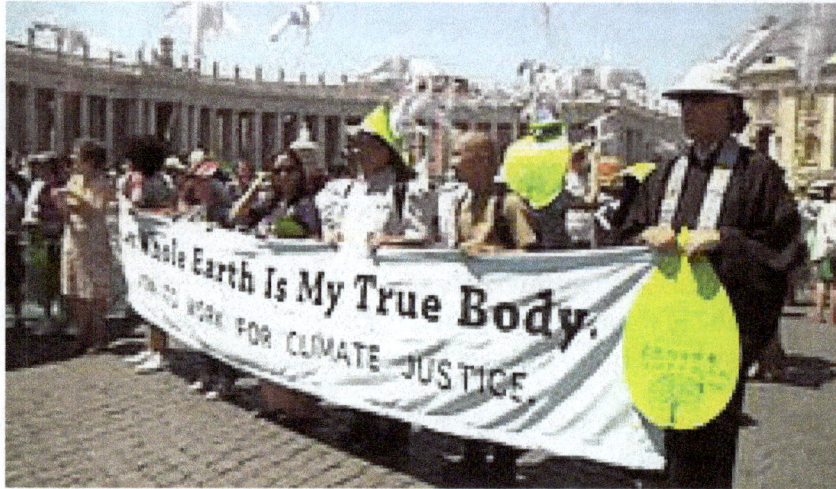

Play Media

Pope Francis's environmental encyclical Laudato si' has been welcomed by many environmental organisations of different faiths - Interfaith march in Rome to call for climate action

The Christian world view sees the universe as created by God, and humankind accountable to God for the use of the resources entrusted to humankind. Ultimate values are seen in the light of being valuable to God. This applies both in breadth of scope - caring for people (Matthew 25) and environmental issues, e.g. environmental health (Deuteronomy 22.8; 23.12-14) - and dynamic motivation, the love of Christ controlling (2 Corinthians 5.14f) and dealing with the underlying spiritual disease of sin, which shows itself in selfishness and thoughtlessness. In many countries this relationship of accountability is symbolised at harvest thanksgiving. (B.T. Adeney : Global Ethics in New Dictionary of Christian Ethics and Pastoral Theology 1995 Leicester)

Abrahamic religious scholars have used theology to motivate the public. John L. O'Sullivan, who coined the term Manifest destiny, and other influential people like him used Abrahamic ideologies to encourage action. These religious scholars, columnists and politicians historically have used these ideas and continue to do so to justify the consumptive tendencies of a young America around the time of the Industrial Revolution. In order to solidify the understanding that God had intended for humankind to use earths natural resources, environmental writers and religious scholars alike proclaimed that humans are separate from nature, on a higher order. Those that may critique this point of view may ask the same question that John Muir asks ironically in a section of his novel *A Thousand Mile Walk to the Gulf, why are there so many dangers in the natural world in the form of poisonous plants, animals and natural disasters,* The answer is that those creatures are a result of Adam and Eve's sins in the garden of Eden.

Since the turn of the 20th century, the application of theology in environmentalism diverged into two schools of thought. The first system of understanding holds religion as the basis of environmental stewardship. The second sees the use of theology as a means to rationalize the unmanaged consumptions of natural resources. Lynn White and Calvin DeWitt represent each side of this dichotomy.

John Muir personified nature as an inviting place away from the loudness of urban centers. "For Muir and the growing number of Americans who shared his views, Satan's home had become God's Own Temple." The use of Abrahamic religious allusions assisted Muir and the Sierra Club to create support for some of the first public nature preserves.

Authors like Terry Tempest Williams as well as John Muir build on the idea that "...God can be found wherever you are, especially outside. Family worship was not just relegated to Sunday in a chapel." References like these assist the general public to make a connection between paintings done at the Hudson River School, Ansel Adam's photographs, along with other types of media, and their religion or spirituality. Placing intrinsic value upon nature through theology is a fundamental idea of Deep Ecology.

Anthropocentrism

Anthropocentrism is the position that humans are the most important or critical element in any given situation; that the human race must always be its own primary concern. Detractors of anthropocentrism argue that the Western tradition biases homo sapiens when considering the environmental ethics of a situation and that humans evaluate they environment or other organisms in terms of the utility for them(see speciesism). Many argue that all environmental studies should include an assessment of the intrinsic value of non-human beings. In fact, based on this very assumption, a philosophical article has explored recently the possibility of humans' willing extinction as a gesture toward other beings. The authors refer to the idea as a thought experiment that should not be understood as a call for action.

What anthropocentric theories do not allow for is the fact that a system of ethics formulated from a human perspective may not be entirely accurate; humans are not necessarily the centre of reality. The philosopher Baruch Spinoza argued that humans tend to assess things wrongly in terms of their usefulness to us. Spinoza reasoned that if humans were to look at things objectively, they would discover that everything in the universe has a unique value. Likewise, it is possible that a human-centred or anthropocentric/androcentric ethic is not an accurate depiction of reality, and there is a bigger picture that humans may or may not be able to understand from a human perspective.

Peter Vardy distinguished between two types of anthropocentrism. A strong anthropocentric ethic argues that humans are at the center of reality and it is right for them to be so. Weak anthropocentrism, however, argues that reality can only be interpreted from a human point of view, thus humans have to be at the centre of reality as they see it.

Another point of view has been developed by Bryan Norton, who has become one of the essential actors of environmental ethics by launching environmental pragmatism, now one of its leading trends. Environmental pragmatism refuses to take a stance in disputes between defenders of anthropocentrist and non-anthropocentrist ethics. Instead, Norton distinguishes between *strong anthropocentrism* and *weak-or-extended-anthropocentrism* and argues that the former must underestimate the diversity of instrumental values humans may derive from the natural world.

A recent view relates anthropocentrism to the future of life. Biotic ethics are based on the human identity as part of gene/protein organic life whose effective purpose is self-propagation. This implies a human purpose to secure and propagate life. Humans are central because only they can secure life beyond the duration of the Sun, possibly for trillions of eons. Biotic ethics values life itself, as embodied in biological structures and processes. Humans are special because they can secure the future of life on cosmological scales. In particular, humans can continue sentient life that enjoys its existence, adding further motivation to propagate life. Humans can secure the future of life, and this future can give human existence a cosmic purpose.

Status of the Field

Only after 1990 did the field gain institutional recognition at programs such as Colorado State University, the University of Montana, Bowling Green State University, and the University of North Texas. In 1991, Schumacher College of Dartington, England, was founded and now provides an MSc in Holistic Science.

These programs began to offer a master's degree with a specialty in environmental ethics/philosophy. Beginning in 2005 the Department of Philosophy and Religion Studies at the University of North Texas offered a PhD program with a concentration in environmental ethics/philosophy.

In Germany, the University of Greifswald has recently established an international program in Landscape Ecology & Nature Conservation with a strong focus on environmental ethics. In 2009, the University of Munich and Deutsches Museum founded the Rachel Carson Center for Environment and Society, an international, interdisciplinary center for research and education in the environmental humanities.

Indigenous Rights

On 26 August 1975 Prime Minister Gough Whitlam handed a leasehold title to land at Daguragu (Wattie Creek) to Vincent Lingiari, representative of the Gurindji people.

Indigenous rights are those rights that exist in recognition of the specific condition of the indigenous peoples. This includes not only the most basic human rights of physical survival and integrity, but also the preservation of their land, language, religion, and other elements of cultural heritage that are a part of their existence as a people. This can be used as an expression for advocacy of social organizations or form a part of the national law in establishing the relation between a government and the right of self-determination among the indigenous people living within its borders, or in international law as a protection against violation by actions of governments or groups of private interests.

Definition and Historical Background

The indigenous rights belong to those who, being indigenous peoples, are defined by being the original people of a land that has been conquested and colonized by outsiders. Exactly who is a part of the indigenous peoples is disputed, but can broadly be understood in relation to colonialism. When we speak of indigenous peoples we speak of those pre-colonial societies that face a specific threat from this phenomenon of occupation, and the relation that these societies have with the colonial powers. The exact definition of who are the indigenous people, and the consequent state of rightsholders, varies. It is considered both to be bad to be too inclusive as it is to be non-inclusive. In the context of modern indigenous people of European colonial powers, the recognition of indigenous rights can be traced to at least the period of Renaissance. Along with the justification of colonialism with a higher purpose for both the colonists and colonized, some voices expressed concern over the way indigenous peoples were treated and the effect it had on their societies.

The issue of indigenous rights is also associated with other levels of human struggle. Due to the close relationship between indigenous peoples' cultural and economic situations and their environmental settings, indigenous rights issues are linked with concerns over environmental change and sustainable development. According to scientists and organizations like the Rainforest Foundation, the struggle for indigenous peoples is essential for solving the problem of reducing carbon emission, and approaching the threat on both cultural and biological diversity in general.

Representation

The rights, claims and even identity of indigenous peoples are apprehended, acknowledged and observed quite differently from government to government. Various organizations exist with charters to in one way or another promote (or at least acknowledge) indigenous aspirations, and indigenous societies have often banded together to form bodies which jointly seek to further their communal interests.

International Organizations

There are several non-governmental civil society movements, networks, indigenous and non-indigenous organizations, such as International Indian Treaty Council, Indigenous World Association, the International Land Coalition, ECOTERRA Intl. , Indigenous Environmental Network, Earth Peoples, Global Forest Coalition, Amnesty International, Indigenous Peoples Council on Biocolonialism, Friends of Peoples Close to Nature, Indigenous Peoples Issues and Resources, Minority Rights Group International, Survival International and Cultural Survival, whose founding mission is to protect indigenous rights, including land rights. These organizations, networks and groups underline that the problems that indigenous peoples are facing is the lack of recognition that they are entitled to live the way they choose, and lack of the right to their lands and territories. Their mission is to protect the rights of indigenous peoples without states imposing their ideas of "development". These groups say that each indigenous culture is differentiated, rich of religious believe systems, way of life, substenance and arts, and that the root of problem would be the interference with their way of living by state's disrespect to their rights, as well as the invasion of traditional lands by multinational cooperations and small businesses for exploitation of natural resources.

United Nations

Indigenous peoples and their interests are represented in the United Nations primarily through the mechanisms of the Working Group on Indigenous Populations (WGIP). In April 2000 the United Nations Commission on Human Rights adopted a resolution to establish the United Nations Permanent Forum on Indigenous Issues (PFII) as an advisory body to the Economic and Social Council with a mandate to review indigenous issues.

In late December 2004, the United Nations General Assembly proclaimed 2005–2014 to be the *Second International Decade of the World's Indigenous People*. The main goal of the new decade will be to strengthen international cooperation around resolving the problems faced by indigenous peoples in areas such as culture, education, health, human rights, the environment, and social and economic development.

In September 2007, after a process of preparations, discussions and negotiations stretching back to 1982, the General Assembly adopted the Declaration on the Rights of Indigenous Peoples. The non-binding declaration outlines the individual and collective rights of indigenous peoples, as well as their rights to identity, culture, language, employment, health, education and other issues. Four nations with significant indigenous populations voted against the declaration: the United States, Canada, New Zealand and Australia. All four have since then changed their vote in favour. Eleven nations abstained: Azerbaijan,Bangladesh, Bhutan, Burundi, Colombia, Georgia, Kenya, Nigeria, Russia, Samoa and Ukraine. Thirty-four nations did not vote, while the remaining 143 nations voted for it.

ILO 169

ILO 169 is a convention of the International Labour Organisation. Once ratified by a state, it is meant to work as a law protecting tribal people's rights. There are twenty-two physical survival and integrity, but also the preservation of their land, language and religion rights. The ILO is represents indigenous rights as they are the organisation that enforced instruments the deal with indigenous rights exclusively.

Definition and Historical Background

The indigenous rights belong to those who, being indigenous peoples, are defined by being the original people of a land that has been invaded and colonizedby outsiders. Exactly who is a part of the indigenous peoples is disputed, but can broadly be understood in relation to colonialism. When we speak of indigenous peoples we speak of those pre-colonial societies that face a specific threat from this phenomenon of occupation, and the relation that these societies have with the colonial powers. The exact definition of who are the indigenous people, and the consequent state of rightsholders, varies. It is considered both to be bad to be too inclusive as it is to be non-inclusive. In the context of modern indigenous people of European colonial powers, the recognition of indigenous rights can be traced to at least the period of Renaissance. Along with the justification of colonialism with a higher purpose for both the colonists and colonized, some voices ountries that ratified the *Convention 169* since the year of adoption in 1989: Argentina, Bolivia, Brazil, Central African Republic, Chile, Colombia, Costa Rica, Denmark, Dominica, Ecuador, Fiji, Guatemala, Honduras, México, Nepal, Netherlands, Nicaragua, Norway, Paraguay, Peru, Spain and Venezuela. The law recognizes land ownership; equality and freedom; and autonomy for decisions affecting indigenous peoples.

Organization of American States

Since 1997, the nations of the Organization of American States have been discussing draft versions of a proposed American Declaration on the Rights of Indigenous Peoples. "The draft declaration is currently one of the most important processes underway with regard to indigenous rights in the Americas"as mentioned by the International Work Group for Indigenous Affairs.

Protected Area

Protected areas or conservation areas are locations which receive protection because of their recognized natural, ecological and/or cultural values. There are several kinds of protected areas, which vary by level of protection depending on the enabling laws of each country or the regulations of the international organisations involved. The term "protected area" also includes Marine Protected Areas, the boundaries of which will include some area of ocean, and Transboundary Protected Areas that overlap multiple countries which remove the borders inside the area for conservation and economic purposes. There are over 161,000 protected areas in the world (as of October 2010) with more added daily, representing between 10 and 15 percent of the world's land surface area. By contrast, only 1.17% of the world's oceans is included in the world's ~6,800 Marine Protected Areas.

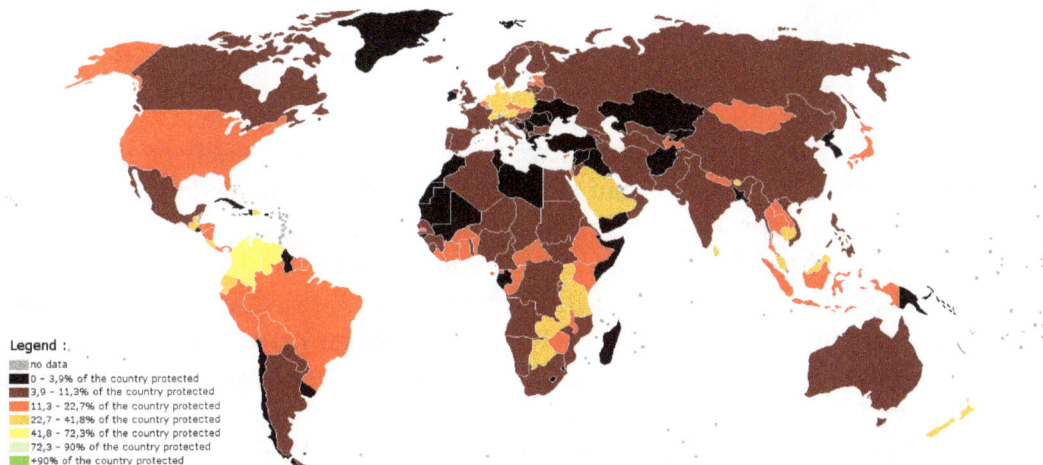

Legend :
no data
0 – 3,9% of the country protected
3,9 – 11,3% of the country protected
11,3 – 22,7% of the country protected
22,7 – 41,8% of the country protected
41,8 – 72,3% of the country protected
72,3 – 90% of the country protected
+90% of the country protected

World map with total percentage of each country under protection.

Protected areas are essential for biodiversity conservation, often providing habitat and protection from hunting for threatened and endangered species. Protection helps maintain ecological processes that cannot survive in most intensely managed landscapes and seascapes.

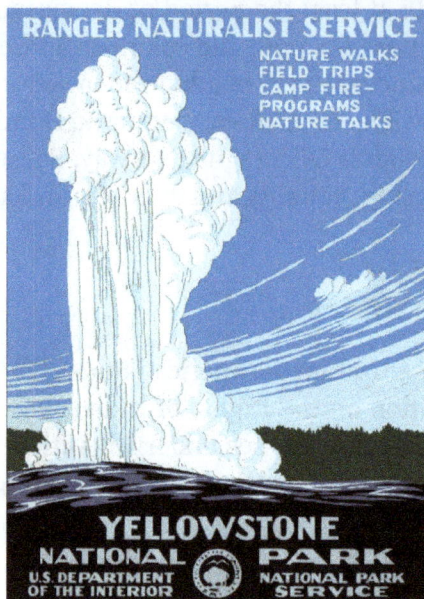

A poster of Yellowstone from 1938.

Definition

Generally, protected areas are understood to be those in which human occupation or at least the exploitation of resources is limited. The definition that has been widely accepted across regional and global frameworks has been provided by the International Union for Conservation of Nature (IUCN) in its categorisation guidelines for protected areas. The definition is as follows:

"A clearly defined geographical space, recognized, dedicated and managed, through legal or other effective means, to achieve the long-term conservation of nature with associated ecosystem services and cultural values."

Protection of Natural Resources

Protected areas are designated with the objective of conserving biodiversity and providing an indicator for that conservation's progress, but the extent to which they defend resources and ecosystem dynamics from degradation are slightly more complex. Protected areas will usually encompass several other zones that have been deemed important for particular conservation uses, such as Important Bird Areas (IBA) and Endemic Bird Areas (EBA), Centres of Plant Diversity (CBD), Indigenous and Community Conserved Areas (ICCA), Alliance for Zero Extinction Sites (AZE) and Key Biodiversity Areas (KBA) among others. Likewise, a protected area or an entire network of protected areas may lie within a larger geographic zone that is recognised as a terrestrial or marine ecoregions (see, Global 200), or a crisis ecoregions for example. As a result, Protected Areas can encompass a broad range of governance types. Indeed, governance of protected areas has emerged a critical factor in their success.

Subsequently, the range of natural resources that any one protected area may guard is vast. Many will be allocated primarily for species conservation whether it be flora or fauna or the relationship between them, but protected areas are similarly important for conserving sites of (indigenous) cultural importance and considerable reserves of natural resources such as;

- Carbon Stocks: Carbon emissions from deforestation account for an estimated 20% of global carbon emissions, so in protecting the worlds carbon stocks greenhouse gas emissions are reduced and longterm land cover change is prevented, which is an effective strategy in the struggle against global warming. Of all global terrestrial carbon stock, 15.2% is contained within protected areas. Protected areas in South America hold 27% of the world's carbon stock, which is the highest percentage of any country in both absolute terms and as a proportion of the total stock.

- Rainforests: 18.8% of the world's forest is covered by protected areas and sixteen of the twenty forest types have 10% or more protected area coverage. Of the 670 ecoregions with forest cover, 54% have 10% or more of their forest cover protected under IUCN Categories I – VI.

- Mountains: Nationally designated protected areas cover 14.3% of the world's mountain areas, and these mountainous protected areas made up 32.5% of the world's total terrestrial protected area coverage in 2009. Mountain protected area coverage has increased globally by 21% since 1990 and out of the 198 countries with mountain areas, 43.9% still have less than 10% of their mountain areas protected.

Annual updates on each of these analyses are made in order to make comparisons to the Millennium Development Goals and several other fields of analysis are expected to be introduced in the monitoring of protected areas management effectiveness, such as freshwater and marine or coastal studies which are currently underway, and islands and drylands which are currently in planning.

IUCN Protected Area Management Categories

Through its World Commission on Protected Areas (WCPA), the IUCN has developed six Protected Area Management Categories that define protected areas according to their management objectives, which are internationally recognised by various national governments and the United Nations. The categories provide international standards for defining protected areas and encourage conservation planning according to their management aims.

Strict Nature reserve Belianske Tatras in Slovakia.

IUCN Protected Area Management Categories:

- Category Ia — Strict Nature Reserve

- Category Ib — Wilderness Area

- Category II — National Park

- Category III — Natural Monument or Feature

- Category IV — Habitat/Species Management Area

- Category V — Protected Landscape/Seascape

- Category VI – Protected Area with sustainable use of natural resources

History

Protected areas are cultural artifacts, and their story is entwined with that of human civilization. Protecting places and resources is by no means a modern concept, whether it be indigenous communities guarding sacred sites or the convention of European hunting reserves. Over 2000 years ago, royal decrees in India protected certain areas. In Europe, rich and powerful people protected hunting grounds for a thousand years. Moreover, the idea of protection of special places is universal: for example, it occurs among the communities in the Pacific ("tapu" areas) and in parts of Africa (sacred groves).

Black Opal Spring in Yellowstone National Park in the United States. Yellowstone, the world's second official protected area (after Mongolia's Bogd Khan Mountain), was declared a protected area in 1872, and it encompasses areas which are classified as both a National Park (Category II) and a Habitat Management Area (Category IV).

In 1778 during the reign of the Tenger Tetgegch Khaan, Qing China approved a protected area on then-Khan Uul, a mountain previous protected by local nomads for centuries, in Mongolia, making it the first officially protected area in the world in the present day criteria. However, the mass protected areas movement doesn't begin until late nineteenth-century in North America, Australia, New Zealand and South Africa, when other countries were quick to follow suit. While the idea of protected areas spread around the world in the twentieth century, the driving force was different in different regions. Thus, in North America, protected areas were about safeguarding dramatic and sublime scenery; in Africa, the concern was with game parks; in Europe, landscape protection was more common.

Initially, protected areas were recognised on a national scale, differing from country to country until 1933, when an effort to reach an international consensus on the standards and terminology of protected areas took place at the International Conference for the Protection of Fauna and Flora in London. At the 1962 First World Conference on National Parks in Seattle the effect the Industrial Revolution had had on the world's natural environment was acknowledged, and the need to preserve it for future generations was established.

Since then, it has been an international commitment on behalf of both governments and non-government organisations to maintain the networks that hold regular revisions for the succinct categorisations that have been developed to regulate and record protected areas. In 1972, the Stockholm Declaration of the United Nations Conference on the Human Environment endorsed the protection of representative examples of all major ecosystem types as a fundamental requirement of national conservation programmes. This has become a core principle of conservation biology and has remained so in recent resolutions - including the World Charter for Nature in 1982, the Rio Declaration at the Earth Summit in 1992, and the Johannesburg Declaration 2002.

Recently, the importance of protected areas has been brought to the fore at the threat of human-induced global warming and the understanding of the necessity to consume natural resources in a sustainable manner. The spectrum of benefits and values of protected areas is recognised not only ecologically, but culturally through further development in the arena of Indigenous and Community Conserved Areas (ICCAs). International programmes for the protection of representative ecosystems remain relatively progressive (considering the environmental challenges of globalisation with respect to terrestrial environments), with less advances in marine and freshwater biomes.

Challenges

How to manage areas protected for conservation brings up a range of challenges - whether it be regarding the local population, specific ecosystems or the design of the reserve itself - and because of the many unpredicatable elements in ecology issues, each protected area requires a case-specific set of guidelines.

Schweizerischer National Park in the Swiss Alps is a Strict Nature Reserve (Category Ia).

Enforcing protected area boundaries is a costly and labour-heavy endeavour, particularly if the allocation of a new protected region places new restrictions on the use of resources by the native people which may lead to their subsequent displacement. This has troubled relationships between conservationists and rural communities in many protected regions and is often why many Wildlife

Reserves and National Parks face the human threat of poaching for the illegal bushmeat or trophy trades, which are resorted to as an alternative form of substinence.

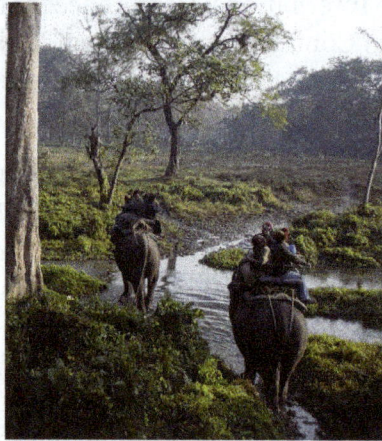

The Jaldapara National Park in West Bengal, India is a Habitat Management Area (Category IV).

There is increasing pressure to take proper account of human needs when setting up protected areas and these sometimes have to be "traded off" against conservation needs. Whereas in the past governments often made decisions about protected areas and informed local people afterwards, today the emphasis is shifting towards greater discussions with stakeholders and joint decisions about how such lands should be set aside and managed. Such negotiations are never easy but usually produce stronger and longer-lasting results for both conservation and people.

In some countries, protected areas can be assigned without the infrastructure and networking needed to substitute consumable resources and subtantiatively protect the area from development or misuse. The soliciting of protected areas may require regulation to the level of meeting demands for food, feed, livestock and fuel, and the legal enforcement of not only the protected area itself but also 'buffer zones' surrounding it, which may help to resist destabilisation.

Effectiveness

One of the main concerns regarding protected areas on land and sea is their effectiveness at preventing the ongoing loss of biodiversity. There are multiple case studies indicating the positive effects of protected areas on terrestrial and marine species. However, those cases do not represent the majority of protected areas. Several limitations that may preclude their success include: their small size and large isolation to each other (both of these factors influence the maintenance of species), their limited role at preventing the many factors affecting biodiversity (e.g. climate change, invasive species, pollution), their large cost and their increasing conflict with human demands for nature's goods and services.

By Area

European Union

Natura 2000 is a network of protected areas established by the European Union across all Member States. It is made up of Special Areas of Conservation (SACs) and Special Protection Areas (SPAs) designated respectively under the Habitats Directive and Birds Directive. 787,767 km^2

(304,159 sq mi) are designated as terrestrial sites and 251,564 km² (97,129 sq mi) as marine sites. Overall, 18 percent of the EU land mass is designated.

Nicaragua

O Parks, Wildlife, and Recreation is a Private Protected Area, also known as a 'Private Reserve' predominantly managed for biodiversity conservation, protected without formal government recognition and is owned and stewarded by the O corporation International. O parks plays a particularly important role in conserving critical biodiversity in a section of the Mesoamerican Biological Corridor known as the Paso del Istmo, located along the 12-mile-wide isthmus between Lake Nicaragua and the Pacific Ocean.

United States

As of 31 January 2008, according to the United Nations Environment Programme, the United States had a total of 6,770 terrestrial nationally designated (federal) protected areas. These protected areas cover 2,607,131 km² (1,006,619 sq mi), or 27.08 percent of the land area of the United States. This is also one-tenth of the protected land area of the world.

Prohibited activities and safety instructions at a state park in Oregon

Various Protected Areas

Marine Protected Area

Marine protected areas (MPA) are protected areas of seas, oceans, estuaries or large lakes. MPAs restrict human activity for a conservation purpose, typically to protect natural or cultural resources. Such marine resources are protected by local, state, territorial, native, regional, national, or international authorities and differ substantially among and between nations. This variation includes different limitations on development, fishing practices, fishing seasons and catch limits, moorings and bans on removing or disrupting marine life. In some situations (such as with the

Phoenix Islands Protected Area), MPAs also provide revenue for countries, potentially equal to the income that they would have if they were to grant companies permissions to fish.

Milford Sound, New Zealand is a *strict marine reserve* (Category Ia) Mitre Peak, the mountain at left, rises 1,692 m (5,551 ft) above the sea.

On 28 October 2016 in Hobart, Australia, the Convention for the Conservation of Antarctic Marine Living Resources agreed to establish the first Antarctic and largest marine park in the world encompassing 1.55 million km² (600,000 sq mi) in the Ross Sea. Other large MPAs are in the Indian, Pacific, and Atlantic Oceans in certain exclusive economic zones of Australia and overseas territories of France, the United Kingdom and the United States, with major (990,000 square kilometres (380,000 sq mi) or larger) new or expanded MPAs by these nations since 2012—such as Natural Park of the Coral Sea, Pacific Remote Islands Marine National Monument, Coral Sea Commonwealth Marine Reserve and South Georgia and the South Sandwich Islands Marine Protected Area. When counted with MPAs of all sizes from many other countries, as of August 2016 there are more than 13,650 MPAs, encompassing 2.07% of the world's oceans, with half of that area – encompassing 1.03% of the world's oceans – receiving complete "no-take" designation.

Terminology

"MPA" is an umbrella term for protected areas that includes some area of marine landscape and/or biodiversity. The IUCN defines a marine protected area as:

"Any area of the intertidal or subtidal terrain, together with its overlying water and associated flora, fauna, historical and cultural features, which has been reserved by law or other effective means to protect part or all of the enclosed environment."

An alternative is "a clearly defined geographical space, recognized, dedicated, and managed through legal or other effective means, to achieve the long term conservation of nature with associated ecosystem services and cultural values". United States Executive Order 13158 in May 2000 established MPAs, defining them as;

"Any area of the marine environment that has been reserved by federal, state, tribal, territorial, or local laws or regulations to provide lasting protection for part or all of the natural and cultural resources therein."

The Convention on Biological Diversity defined the broader term of *marine and coastal protected area* (MCPA);

"Any defined area within or adjacent to the marine environment, together with its overlying water and associated flora, fauna, historical and cultural features, which has been reserved by legislation or other effective means, including custom, with the effect that its marine and/or coastal biodiversity enjoys a higher level of protection than its surroundings."

Classifications

The Chagos Archipelago was declared the world's largest marine reserve in April 2010 with an area of 250,000 square miles until March 2015 when It was declared illegal by the Permanent Court of Arbitration.

Several types of compliant MPA can be distinguished:

- A totally marine area with no significant terrestrial parts.

- An area containing both marine and terrestrial components, which can vary between two extremes; those that are predominantly maritime with little land (for example, an atoll would have a tiny island with a significant maritime population surrounding it), or that is mostly terrestrial.

- Marine ecosystems that contain land and intertidal components only. For example, a mangrove forest would contain no open sea or ocean marine environment, but its river-like marine ecosystem nevertheless complies with the definition.

IUCN offered seven categories of protected area, based on management objectives and four broad governance types.

Cat	IUCN Protected Area Management Categories:
Ia	Strict nature reserve A marine reserve usually connotes "maximum protection", where all resource removals are strictly prohibited. In countries such as Kenya and Belize, marine reserves allow for low-risk removals to sustain local communities.
Ib	Wilderness area
II	National park Marine parks emphasize the protection of ecosystems but allow light human use. A marine park may prohibit fishing or extraction of resources, but allow recreation. Some marine parks, such as those in Tanzania, are zoned and allow activities such as fishing only in low risk areas.

III	Natural monuments or features Established to protect historical sites such as shipwrecks and cultural sites such as aboriginal fishing grounds.
IV	Habitat/species management area Established to protect a certain species, to benefit fisheries, rare habitat, as spawning/nursing grounds for fish, or to protect entire ecosystems.
V	Protected seascape Limited active management, as with protected landscapes.
VI	Sustainable use of natural resources

Related protected area categories include the following;

- World Heritage Site (WHS) – an area exhibiting extensive natural or cultural history. Maritime areas are poorly represented, however, with only 46 out of over 800 sites.

- Man and the Biosphere – UNESCO program that promotes "a balanced relationship between humans and the biosphere". Under article 4, biosphere reserves must "encompass a mosaic of ecological systems", and thus combine terrestrial, coastal, or marine ecosystems. In structure they are similar to Multiple-use MPAs, with a core area ringed by different degrees of protection.

- Ramsar site – must meet certain criteria for the definition of "Wetland" to become part of a global system. These sites do not necessarily receive protection, but are indexed by importance for later recommendation to an agency that could designate it a protected area.

While "area" refers to a single contiguous location, terms such as *"network"*, *"system"*, and *"region"* that group MPAs are not always consistently employed.*"System"* is more often used to refer to an individual MPA, whereas *"region"* is defined by the World Conservation Monitoring Centre as:

"A collection of individual MPAs operating cooperatively, at various spatial scales and with a range of protection levels that are designed to meet objectives that a single reserve cannot achieve."

At the 2004 Convention on Biological Diversity, the agency agreed to use *"network"* on a global level, while adopting *system* for national and regional levels. The *network* is a mechanism to establish regional and local systems, but carries no authority or mandate, leaving all activity within the *"system"*.

No take zones (NTZs), are areas designated in a number of the world's MPAs, where all forms of exploitation are prohibited and severely limits human activities. These no take zones can cover an entire MPA, or specific portions. For example, the 1,150,000 square kilometres (440,000 sq mi) Papahānaumokuākea Marine National Monument, the world's largest MPA (and largest protected area of any type, land or sea), is a 100% no take zone.

Related terms include; *specially protected area* (SPA), *Special Area of Conservation* (SAC), the United Kingdom's *marine conservation zones* (MCZs), or *area of special conservation* (ASC) etc. which each provide specific restrictions.

Stressors

Stressors that affect oceans include "the impact of extractive industries, localised pollution, and changes to its chemistry (ocean acidification) resulting from elevated carbon dioxide levels, due to our emissions". MPAs have been cited as the ocean's single greatest hope for increasing the resilience of the marine environment to such stressors. Well-designed and managed MPAs developed with input and support from interested stakeholders can conserve biodiversity and protect and restore fisheries.

Economics

MPAs can help sustain local economies by supporting fisheries and tourism. For example, Apo Island in the Philippines made protected one quarter of their reef, allowing fish to recover, jump-starting their economy. This was shown in the film, *Resources at Risk: Philippine Coral Reef*. A 2016 report by the Center for Development and Strategy found that programs like the United States National Marine Sanctuary system can develop considerable economic benefits for communities through Public–private partnerships.

Management

Typical MPAs restrict fishing, oil and gas mining and/or tourism. Other restrictions may limit the use of ultrasonic devices like sonar (which may confuse the guidance system of cetaceans), development, construction and the like. Some fishing restrictions include "no-take" zones, which means that no fishing is allowed. Less than 1% of US MPAs are no-take.

Ship transit can also be restricted or banned, either as a preventive measure or to avoid direct disturbance to individual species. The degree to which environmental regulations affect shipping varies according to whether MPAs are located in territorial waters, exclusive economic zones, or the high seas. The law of the sea regulates these limits.

Most MPAs have been located in territorial waters, where the appropriate government can enforce them. However, MPAs have been established in exclusive economic zones and in international waters. For example, Italy, France and Monaco in 1999 jointly established a cetacean sanctuary in the Ligurian Sea named the Pelagos Sanctuary for Mediterranean Marine Mammals. This sanctuary includes both national and international waters. Both the CBD and IUCN recommended a variety of management systems for use in a protected area system. They advocated that MPAs be seen as one of many "nodes" in a network of protected areas. The following are the most common management systems:

Asinara, Italy is listed by WDPA as both a marine reserve and a national marine park, and as such could be labelled 'multiple-use'

Seasonal and temporary management—Activities, most critically fishing, are restricted seasonally or temporarily, e.g., to protect spawning/nursing grounds or to let a rapidly reducing species recover.

Multiple-use MPAs—These are the most common and arguably the most effective. These areas employ two or more protections. The most important sections get the highest protection, such as a no take zone and are surrounded with areas of lesser protections.

Community involvement and related approaches—Community-managed MPAs empower local communities to operate partially or completely independent of the governmental jurisdictions they occupy. Empowering communities to manage resources can lower conflict levels and enlist the support of diverse groups that rely on the resource such as subsistence and commercial fishers, scientists, recreation, tourism businesses, youths and others.

MPA networks—"A group of MPAs that interact with one another ecologically and/or socially form a network". These networks are intended to connect individuals and MPAs and promote education and cooperation among various administrations and user groups. "MPA networks are, from the perspective of resource users, intended to address both environmental and socio-economic needs, complementary ecological and social goals and designs need greater research and policy support". Filipino communities connect with one another to share information about MPAs, creating a larger network through the social communities' support. Emerging or established MPA networks can be found in Southeast Australia, Belize, the Red Sea, Gulf of Aden and Mexico.

International Efforts

The 17th International Union for Conservation of Nature (IUCN) General Assembly in San Jose, California, the 19th IUCN assembly and the fourth World Parks Congress all proposed to centralise the establishment of protected areas. The World Summit on Sustainable Development in 2002 called for

the establishment of marine protected areas consistent with international laws and based on scientific information, including representative networks by 2012.

The Evian agreement, signed by G8 Nations in 2003, agreed to these terms. The Durban Action Plan, developed in 2003, called for regional action and targets to establish a network of protected areas by 2010 within the jurisdiction of regional environmental protocols.It recommended establishing protected areas for 20 to 30% of the world's oceans by the goal date of 2012. The Convention on Biological Diversity considered these recommendations and recommended requiring countries to set up marine parks controlled by a central organization before merging them. The United Nations Framework Convention on Climate Change agreed to the terms laid out by the convention, and in 2004, its member nations committed to the following targets;

- By 2006 complete an area system gap analysis at national and regional levels.

- By 2008 address the less represented marine ecosystems, accounting for those beyond national jurisdiction in accordance.

- By 2009 designate the protected areas identified through the gap analysis.

- By 2012 complete the establishment of a comprehensive and ecologically representative network.

"The establishment by 2010 of terrestrial and by 2012 for marine areas of comprehensive, effectively managed, and ecologically representative national and regional systems of protected areas that collectively, inter alia through a global network, contribute to achieving the three objectives of the Convention and the 2010 target to significantly reduce the current late of biodiversity loss at the global, regional, national, and sub-national levels and contribute to poverty reduction and the pursuit of sustainable development."

The UN later endorsed another decision, Decision VII/15, in 2006:

Effective conservation of 10% of each of the world's ecological regions by 2010. – United Nations Framework Convention on Climate Change Decision VII/15

Global Treaties

United Nations Convention on The Law of The Sea

The Antarctic Treaty System

On 7 April 1982, the Convention on the Conservation of Antarctic Marine Living Resources (CAMLR Convention) came into force after discussions began in 1975 between parties of the then-current Antarctic Treaty to limit large-scale exploitation of krill by commercial fisheries. The Convention bound contracting nations to abide by previously agreed upon Antarctic territorial claims and peaceful use of the region while protecting ecosystem integrity south of the Antarctic Convergence and 60 S latitude. In so doing, it also established a commission of the original signatories and acceding parties called the Commission for the Conservation of Antarctic Marine Living Resources (CCAMLR) to advance these aims through protection, scientific study, and rational use, such as harvesting, of those marine resources. Though separate, the Antarctic Treaty and CCAMLR, make up part the broader system of international agreements called the Antarctic Treaty System. Since 1982, the CCAMLR meets annually to implement binding conservations measures like the creation of 'protected areas' at the suggestion of the convention's scientific committee.

In 2009, the CCAMLR created the first 'high-seas' MPA entirely within international waters over the southern shelf of the South Orkney Islands. This area encompasses 94,000 square kilometres (36,000 sq mi) and all fishing activity including transhipment, and dumping or discharge of waste is prohibited with the exception of scientific research endeavors. On 28 October 2016, the CCAMLR, composed of 24 member countries and the European Union at the time, agreed to establish the world's largest marine park encompassing 1.55 million km² (600,000 sq mi) in the Ross Sea after several years of failed negotiations. Establishment of the Ross Sea MPA required unanimity of the commission members and enforcement will begin in December 2017. However, due to a sunset provision inserted into the proposal, the new marine park will only be in force for 35 years.

Regional Organizations

PIMPAC

NAMPAM

CCAMLR

National Targets

Many countries have established national targets, accompanied by action plans and implementations. The UN Council identified the need for countries to collaborate with each other to establish effective regional conservation plans. Some national targets are listed in the table below

Country	Plan of action
American Samoa	20% of reefs to be protected by 2010
Australia – South Australia	19 marine protected areas by 2010
Bahamas	20% of the marine ecosystem protected for fishery replenishment by 2010. 20% of coastal and marine habitats by 2015.
Belize	20% of bioregions. 30% of Coral reefs. 60% of turtle nesting sites. 30% of Manatee distribution. 60% of American crocodile nesting. 80% of breeding areas.
Chile	10% of marine areas by 2010. National network for organization by 2015.
Cuba	22% of land habitat, including: 15% of the insular shelf 25% of coral reefs 25% of wetlands
Dominican Republic	20% of marine and coastal by 2020.
Micronesia	30% of shoreline ecosystems by 2020.
Fiji	30% of reefs by 2015. 30% of water managed by marine protected areas by 2020.
Germany	38% of water managed by the marine protected network. (no set date)
Grenada	25% of nearby marine resources by 2020.
Guam	30% of nearby marine ecosystem by 2020.
Indonesia	100,000 km² by 2010. 200,000 km² by 2020.
Ireland	14% of territorial waters as of 2009
Jamaica	20% of marine habitats by 2020.
Madagascar	100,000 km² by 2012.

Marshal Islands	30% of nearby marine ecosystem by 2020.
New Zealand	20% of marine environment by 2010.
North Mariana Islands	30% of nearby marine ecosystem by 2020.
Palau	30% of nearby marine ecosystem by 2020.
Peru	Marine protected area system established by 2015.
Philippines	10% fully protected by 2020.
Senegal	Creation of MPA network. (no set date)
St. Vincent and the Grenadines	20% of marine areas by 2020.
Tanzania	10% of marine area by 2010; 20% by 2020.
United Kingdom	Establish an ecologically coherent network of marine protected areas by 2012.
United States – California	29 MPAs covering 18% of state marine area with 243 square kilometres (94 sq mi) at maximum protection.

National Efforts

The marine protected area network is still in its infancy. As of October 2010, approximately 6,800 MPAs had been established, covering 1.17% of global ocean area. Protected areas covered 2.86% of exclusive economic zones (EEZs). MPAs covered 6.3% of territorial seas. Many prohibit the use of harmful fishing techniques yet only 0.01% of the ocean's area is designated as a "no take zone". This coverage is far below the projected goal of 20%-30% Those targets have been questioned mainly due to the cost of managing protected areas and the conflict that protections have generated with human demand for marine goods and services.

Greater Caribbean

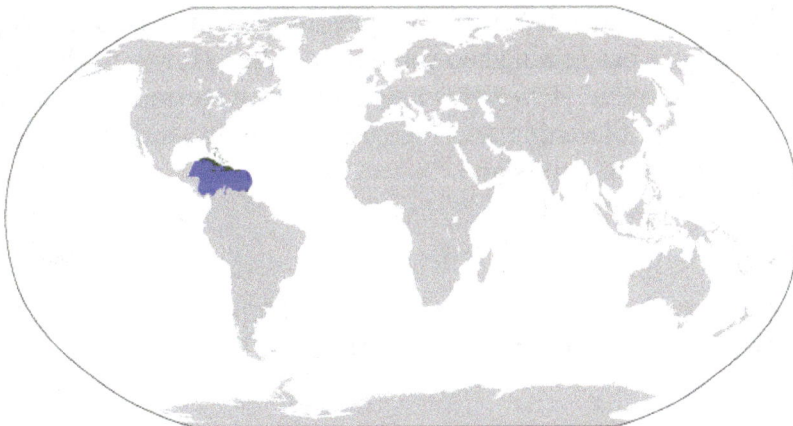

The Caribbean region; the UNEP–defined region also includes the Gulf of Mexico. This region is encompassed by the Mesoamerican Barrier Reef System proposal, and the Caribbean challenge

The Greater Caribbean subdivision encompasses an area of about 5,700,000 square kilometres (2,200,000 sq mi) of ocean and 38 nations. The area includes island countries like the Bahamas and Cuba, and the majority of Central America. The Convention for Protection and Development of the Marine Environment of the Wider Caribbean Region (better known as the Cartagena Convention) was established in 1983. Protocols involving protected areas were ratified in 1990. As of 2008, the region hosted about 500 MPAs. Coral reefs are the best represented.

The Gulf of Mexico region (in 3D) is encompassed by the "Islands in the Stream" proposal.

Two networks are under development, the Mesoamerican Barrier Reef System (a long barrier reef that borders the coast of much of Central America), and the "Islands in the Stream" program (covering the Gulf of Mexico).

Southeast Asia

Southeast Asia is a global epicenter for marine diversity. 12% of its coral reefs are in MPAs. The Philippines have some the world's best coral reefs and protect them to attract international tourism. Most of the Philippines' MPAs are established to secure protection for its coral reef and sea grass habitats. Indonesia has MPAs designed for tourism and relies on tourism as a main source of income.

Philippines

The Philippines host one of the most highly biodiverse regions, with 464 reef-building coral species. Due to overfishing, destructive fishing techniques, and rapid coastal development, these are in rapid decline. The country has established some 600 MPAs. However, the majority are poorly enforced and are highly ineffective. However, some have positively impacted reef health, increased fish biomass, decreased coral bleaching and increased yields in adjacent fisheries. One notable example is the MPA surrounding Apo Island.

Latin America

Latin America has designated one large MPA system. As of 2008, 0.5% of its marine environment was protected, mostly through the use of small, multiple-use MPAs.

South Pacific

The South Pacific network ranges from Belize to Chile. Governments in the region adopted the Lima convention and action plan in 1981. An MPA-specific protocol was ratified in 1989. The permanent commission on the exploitation and conservation on the marine resources of the South Pacific promotes the exchange of studies and information among participants.

The region is currently running one comprehensive cross-national program, the Tropical Eastern Pacific Marine Corridor Network, signed in April 2004. The network covers about 211,000,000 square kilometres (81,000,000 sq mi).

One alternative to imposing MPAs on an indigenous population is through the use of Indigenous Protected Areas, such as those in Australia.

North Pacific

The North Pacific network covers the western coasts of Mexico, Canada, and the U.S. The "Antigua Convention" and an action plan for the north Pacific region were adapted in 2002. Participant nations manage their own national systems. In 2010-2011, the State of California completed hearings and actions via the state Department of Fish and Game to establish new MPAs.

United States and Pacific Island Territories

President Barack Obama signed a proclamation on September 25, 2014, designating the world's largest marine reserve. The proclamation expanded the existing Pacific Remote Islands Marine National Monument, one of the world's most pristine tropical marine environments, to six times its current size, encompassing 490,000 square miles (1,300,000 km²) of protected area around these islands. Expanding the Monument protected the area's unique deep coral reefs and seamounts.

Diagram illustrating the orientation of the 3 marine sanctuaries of Central California: Cordell Bank, Gulf of the Farallones, and Monterey Bay. Davidson Seamount, part of the Monterey Bay sanctuary, is indicated at bottom-right.

In April 2009, the US established a United States National System of Marine Protected Areas, which strengthens the protection of US ocean, coastal and Great Lakes resources. These large-scale MPAs should balance "the interests of conservationists, fishers, and the public." As of 2009, 225 MPAs participated in the national system. Sites work together toward common national and regional conservation goals and priorities. NOAA's national marine protected areas center maintains a comprehensive inventory of all 1,600+ MPAs within the US exclusive economic zone. Most US MPAs.allow some type of extractive use. Fewer than 1% of U.S. waters prohibit all extractive activities.

In 1981 Olympic National Park became a marine protected area. The total protected site area is 3,697 square kilometres (1,427 sq mi). 173.2 km² of the area was an MPA. The national system is a mechanism to foster MPA collaboration. Sites that meet pertinent criteria are eligible to join the national system. Four entry criteria govern admission:

- Meets the definition of an MPA as defined in the Framework.

- Has a management plan (can be sitespecific or part of a broader programmatic management plan; must have goals and objectives and call for monitoring or evaluation of those goals and objectives).

- Contributes to at least one priority conservation objective as listed in the Framework.

- Cultural heritage MPAs must also conform to criteria for the National Register for Historic Places."

In 1999, California adopted the Marine Life Protection Act, establishing the first state law requiring a comprehensive, science-based MPA network. The state created the Marine Life Protection Act Initiative. The MLPA Blue Ribbon Task Force and stakeholder and scientific advisory groups ensure that the process uses the science and public participation.

The MLPA Initiative established a plan to create California's statewide MPA network by 2011 in several steps. The Central Coast step was successfully completed in September, 2007. The North Central Coast step was completed in 2010. The South Coast and North Coast steps were expected to go into effect in 2012.

United Kingdom and British Overseas Territories

United Kingdom

There are a number of marine protected areas around the coastline of the United Kingdom, known as Marine Conservation Zones in England, Wales, and Northern Ireland, Marine Protection Areas in Scotland. They are to be found in inshore and offshore waters.

British Overseas Territories

The United Kingdom is also creating marine protected reserves around several British Overseas Territories. The UK is responsible for 6.8 million square kilometres of ocean around the world, larger than all but four other countries.

The Chagos Marine Protected Area in the Indian Ocean was established in 2010 as a "no-take-zone". With a total surface area of 640,000 square kilometres (250,000 sq mi), it was the world›s largest contiguous marine reserve. In March 2015, the UK announced the creation of a marine reserve around the Pitcairn Islands in the Southern Pacific Ocean to protect its special biodiversity. The area of 830,000 square kilometres (320,000 sq mi) surpassed the Chagos Marine Protected Area as the world's largest contiguous marine reserve, until the August 2016 expansion of the Papahānaumokuākea Marine National Monument in the United States to 1,510,000 square kilometres (580,000 sq mi).

In January 2016, the UK government announced the intention to create a marine protected area around Ascension Island. The protected area will be 234,291 square kilometres (90,460 sq mi), half of which will be closed to fishing.

Europe

The Natura 2000 ecological MPA network in the European Union included MPAs in the North Atlantic, the Mediterranean Sea and the Baltic Sea. The member states had to define NATURA 2000 areas at sea in their Exclusive Economic Zone.

Two assessments, conducted thirty years apart, of three Mediterranean MPAs, demonstrate that proper protection allows commercially valuable and slow-growing red coral (Corallium rubrum) to produce large colonies in shallow water of less than 50 metres (160 ft). Shallow-water colonies outside these decades-old MPAs are typically very small. The MPAs are Banyuls, Carry-le-Rouet and Scandola, off the island of Corsica.

- Mediterranean Science Commission; proposed the creation of 7 marine protected areas ("peace parcs")

Notable Marine Protected Areas

- The Bowie Seamount Marine Protected Area off the coast of British Columbia, Canada.

- The Great Barrier Reef Marine Park in Queensland, Australia.

- The Ligurian Sea Cetacean Sanctuary in the seas of Italy, Monaco and France.

- The Dry Tortugas National Park in the Florida Keys, USA.

- The Papahānaumokuākea Marine National Monument in Hawaii.

- The Phoenix Islands Protected Area, Kiribati.

- The Channel Islands National Marine Sanctuary in California, USA.

- The Chagos Marine Protected Area in the Indian Ocean.

- The Wadden Sea bordering the North Sea in the Netherlands, Germany, and Denmark.

Assessment

Managers and scientists use geographic information systems and remote sensing to map and analyze MPAs. NOAA Coastal Services Center compiled an "Inventory of GIS-Based Decision-Support Tools for MPAs." The report focuses on GIS tools with the highest utility for MPA processes. Remote sensing uses advances in aerial photography image capture, pop-up archival satellite tags, satellite imagery, acoustic data, and radar imagery. Mathematical models that seek to reflect the complexity of the natural setting may assist in planning harvesting strategies and sustaining fishing grounds.

The Prickly Pear Cays are a marine protected area. They're about six miles from Road Bay, Anguilla, in the Leeward Islands of the Caribbean.

Coral Reefs

Coral reef systems have been in decline worldwide. Causes include overfishing, pollution and ocean acidification. As of 2013 30% of the world's reefs were severely damaged. Approximately 60% will be lost by 2030 without enhanced protection. Marine reserves with "no take zones" are the most effective form of protection. Only about 0.01% of the world's coral reefs are inside effective MPAs.

Fish

MPAs can be an effective tool to maintain fish populations. The general concept is to create over-population within the MPA. The fish expand into the surrounding areas to reduce crowding, increasing the population of unprotected areas. This helps support local fisheries in the surrounding area, while maintaining a healthy population within the MPA. Such MPAs are most commonly used for coral reef ecosystems.

One example is at Goat Island Bay in New Zealand, established in 1977. Research gathered at Goat Bay documented the spillover effect. "Spillover and larval export—the drifting of millions of eggs and larvae beyond the reserve—have become central concepts of marine conservation". This positively impacted commercial fishermen in surrounding areas.

Another unexpected result of MPAs is their impact on predatory marine species, which in some conditions can increase in population. When this occurs, prey populations decrease. One study showed that in 21 out of 39 cases, "trophic cascades," caused a decrease in herbivores, which led to an increase in the quantity of plant life. (This occurred in the Malindi Kisite and Watamu Marian National Parks in Kenya; the Leigh Marine Reserve in New Zealand; and Brackett's Landing Conservation Area in the US.

Success Criteria

Both CBD and IUCN have criteria for setting up and maintaining MPA networks, which emphasize 4 factors:

- Adequacy—ensuring that the sites have the size, shape, and distribution to ensure the success of selected species.

- Representability—protection for all of the local environment's biological processes

- Resilience—the resistance of the system to natural disaster, such as a tsunami or flood.

- Connectivity—maintaining population links across nearby MPAs.

Misconceptions

Misconceptions about MPAs include the belief that all MPAs are no-take or no-fishing areas. However, less than 1 percent of US waters are no-take areas. MPA activities can include consumption fishing, diving and other activities.

Another misconception is that most MPAs are federally managed. Instead, MPAs are managed under hundreds of laws and jurisdictions. They can be exist in state, commonwealth, territory and tribal waters.

Another misconception is that a federal mandate dedicates a set percentage of ocean to MPAs. Instead the mandate requires an evaluation of current MPAs and creates a public resource on current MPAs.

Criticism

Some existing and proposed MPAs have been criticized by indigenous populations and their supporters, as impinging on land usage rights. For example, the proposed Chagos Protected Area in the Chagos Islands is contested by Chagossians deported from their homeland in 1965 by the British as part of the creation of the British Indian Ocean Territory (BIOT). According to Wikileaks CableGate documents, the UK proposed that the BIOT become a "marine reserve" with the aim of preventing the former inhabitants from returning to their lands and to protect the joint UK/US military base on Diego Garcia Island.

Other critiques include: their cost (higher than that of passive management), conflicts with human development goals, inadequate scope to address factors such as climate change and invasive species.

Recent Research

The larvae of the yellow tang can drift more than 100 miles and reseed in a distant location.

In 2010, one study found that fish larvae can drift on ocean currents and reseed fish stocks at a distant location. This finding demonstrated that fish populations can be connected to distant locations through the process of larval drift.

They investigated the yellow tang, because larva of this species stay in the general area of the reef in which they first settle. The tropical yellow tang is heavily fished by the aquarium trade. By the late 1990s, their stocks were collapsing. Nine MPAs were established off the coast of Hawaii to protect them. Larval drift has helped them establish themselves in different locations, and the fishery is recovering. "We've clearly shown that fish larvae that were spawned inside marine reserves can drift with currents and replenish fished areas long distances away," said coauthor Mark Hixon.

Marine Reserve

A marine reserve is a type of marine protected area that has legal protection against fishing or development. As of 2007 less than 1% of the world's oceans had been set aside in marine reserves. Benefits include increases in the diversity, density, biomass, body size and reproductive potential of fishery and other species within their boundaries.

Cape Rodney-Okakari Point, Goat Island Marine Reserve (Leigh, Warkworth, New Zealand).

As of 2010, scientists had studied more than 150 marine reserves in at least 61 countries and monitored biological changes inside the reserves. The number of species in each study ranged from 1 to 250 and the reserves ranged in size from 0.006 to 800 square kilometers (0.002 to 310 square miles). In 2014, the World Parks Association adopted a target of establishing no-take zones for 30% of each habitat globally.

Design

A review of studies of 34 families (210 species) of coral reef fishes demonstrates that the design of a marine reserve has important implications for its ability to protect habitat and focal species.

Size and Shape

Effective reserves included habitats that support the life history of focal species (e.g. home ranges, nursery grounds, migration corridors and spawning aggregations), and were located to accommodate movement patterns among them.

Movement patterns (home ranges, ontogenetic shifts and spawning migrations) vary among and within species, and are influenced by factors such as size, sex, behaviour, density, habitat characteristics, season, tide and time of day. For example, damselfishes, butterflyfishes and angelfishes travel <0.1–0.5 km, while some sharks and tuna migrate over thousands of kilometres. Larval dispersal distances tend to be <5–15 km, and self-recruitment to new habitat is common.

The review indicated that effective marine reserves are more than twice the size of the home range of focal/target species (in all directions). The presence of effective marine management outside the reserve may allow smaller reserves. Reserve size recommendations apply to the specific habitats of focal species, not the overall size. For example, coral reef species require coral reef habitats rather than open ocean or seagrass beds.

Marine reserve whose boundaries are extensively fished benefit from compact shapes (e.g., squares or circles rather than elongated rectangles). Including whole ecological units (e.g., an offshore reef) can reduce exports where that is desired.

Scale of movement (km) of coral reef and coastal pelagic fish species					
Group	**Species**	**Daily movements: home ranges, territories and core areas of use**	**Ontogenic shifts**	**Spawning**	**Long-term movements of unknown cause**
Anemonefishes	*Amphiprion* spp.	.1			
Angelfishes	some, e.g. *Centropyge* spp.	.1			
	Holocanthus/ Pomacanthus spp.	.5			
Butterflyfishes	*Chaetodon* spp.	.1			
	some, e.g. *Chaetodon striatus*	.5			
Chub	Bermuda sea chub (*Kyphosus sectatrix*)	3			
Damselfishes	most, e.g. *Dascyllus* spp.	.1			
Eel	Indonesian shortfin eel (*Anguilla bicolor bicolor*)		10		
	Moray (*Gymnothorax* spp.)	.1			
Emperor	e.g. *Lethrinus nebulosus*	5			
	Trumpet (*L. miniatus*)				1000s
Filefish	Orange spotted (*Cantherhines pullus*)	.1			
Goatfishes		1			
Groupers	some, including most *Cephalopholis* spp. and *Ephinephelus* spp.	.1			
	Squaretail coralgrouper (*Plectropomus areolatus*)	1			
	some, e.g. *C. sonnerati* and *E. coicoides*	5			
	Gag (*Mycteroperca microlepis*)	20			

	Nassau (*E. striatus*)			100s	
	Leopard coralgrouper (*P. leopardus*)	3	10		
Grunts	e.g. *Haemulon sciurus*	1			
Jobfish	Green jobfish (Aprion virescens)	10			
Kingfishes	*Seriola* spp.	5			
Manta rays	*Manta* spp.	100s			
Marlin/swordfish		1000s			
Parrotfishes	some *Scarus/Sparisoma* spp.	.5			
	Chlorurus spp., ember (*S. rubroviolaceus*)	3			
	Blue-barred (*S. ghobban*)			10	
	Bumphead (*Bolbometopon muricatum*)	10			
	some, e.g. *Scarus rivulatus*	1			
Rabbitfishes	some, e.g. *Siganus lineatus*	1			
Seahorses	*Hippocampus* spp.	.1			
Shark	lemon (*Negaprion* spp.)	5			
	Whitetip reef (*Triaenodon obesus*) and nurse (*Ginglymostoma cirratum*)	10			
	Blacktip reef (*Carcharhinus melanopterus*)	20	100s		
	Galapagos (*C. galapagensis*)	100s			
	Tiger (*Galeocerdo cuvier*)	1000s			
Shoemaker spinefoot	*S. sutor*			5	
Silver drummer	(*Kyphosus sydneyanus*)	5			
Snappers	some, e.g. *Lutjanus carponotatus*	.1			
	some, e.g. *L. ehrenbergii*	.5			
	Red (*L. campechanus*)	5			
	Mangrove red (*L. argentimaculatus*)			1000s	
Soldierfishes/ squirrelfishes	*Holocentrus* spp./ *Myripristis* spp.	.1			
Surgeonfishes	some (e.g. *Acanthurus lineatus*)	.1			

	some, (e.g. *A. coeruleus* and *Ctenochaetus striatus*)	.5			
	some (e.g. *A. blochii*)	5			
	Yellow tang (*Zebrasoma flavescens*)	1			
	Twotone tang (*Z. scopas*)	1			
Sweetlips	Goldspotted (*Plectorhinchus flavomaculatus*)	3			
Trevally	Bigeye (*Caranx sexfasciatus*)		3		1000s
	Giant (*C. ignobilis*)	20	5		1000s
Triggerfish	Grey (*Balistes capriscus*)	20			
Tuna		1000s			
Unicornfish	Bignose (e.g. *Naso vlamingii*)	.1			
	Bluespine (N. unicornis)	1			
	N. lituratus	5			
Wrasses	most (e.g., *Halichoeres garnoti*)	.1			
	some (e.g. Coris aygula)	5			
	Humphead (*Cheilinus undulatus*)	10			

Habitats

Minimum sustainable population sizes have not been determined for most marine populations. Instead, fisheries ecologists use the fraction of unfished stock levels as a proxy. Meta-analyses suggest that maintaining populations above ~37% of those levels generally ensures stable populations, although variations in fishing pressure allow fractions as small as 10% or as large as 40% (to protect species such as sharks and some grouper that have lower reproductive output or slower maturation). Higher fractions of habitat protection may protect areas vulnerable to disturbances such as typhoons or climate change. 20–30% protection can achieve fisheries objectives in areas with controlled fishing pressure and is the minimum level of habitat protection recommended by IUCN-WCPA.

Special Areas

Many fish species congregate to facilitate spawning. Such congregations are spatially and temporally predictable and increase the species' vulnerability to overfishing. Species such as groupers and rabbitfishes travel long distances to congregate for days or weeks. Such gatherings are their only opportunities to reproduce and are crucial to population maintenance. Species such as snappers and parrotfishes congregate in feeding or resting areas. Juveniles may congregate in nursery areas without adults. Such special areas may require only seasonal protections if at other times no vital activities are taking place. Such reserves must be spaced to allow focal species to journey among them. If the location of such special areas is unknown, or is too large to include in a reserve, management approaches such as seasonal capture and sales restrictions may provide some protection. Sea turtle nesting areas, dugong feeding areas, cetacean migratory corridors and calving grounds are examples of other special areas that can be protected seasonally. Other types of special areas include isolated habitats that have unique assemblages and populations, habitats that are important for endemic species and highly diverse areas.

Isolated Populations

Isolated populations (e.g.,those on remote atolls) have high conservation value where they harbor endemic species and/or unique assemblages. A location or population 20–30 km from its nearest neighbor generally qualifies as isolated in the absence of a persistent linking current. Their isolation (low connectivity) requires such areas to be largely self-replenishing. This leaves them less resilient to disturbance. Sustaining their marine species requires a higher fraction of living areas to be protected.

Recovery

Coral reef fish species recovery rates (from e.g., overfishing) depend on their life history and factors such as ecological characteristics, fishing intensity and population size. In the Coral Triangle, species at lower trophic levels that have smaller maximum sizes, faster growth and maturation rates and shorter life spans tend to recover more quickly than species having the opposite characteristics. For example, in the Philippines, populations of planktivores (e.g., fusiliers) and some herbivores (e.g., parrotfishes) recovered in <5–10 years in marine reserves, while predators (e.g., groupers) took 20–40 years. Increased fishing pressure adversely affects recovery rates (e.g., Great Barrier Reef and Papua New Guinea).

Long-term protection allows species with slower recovery rates to achieve and maintain ecosystem health and associated fishery benefits. Permanent protection protects these species over the long-term. Short-term protections do not allow slow-recovering species to reach or maintain stable populations.

In some Coral Triangle countries (e.g., Papua New Guinea and Solomon Islands), short term protections are the most common form of traditional marine resource management. These protetcions can help address problems at lower trophic levels (e.g., herbivores) or allow spawning to succeed. Other reasons for adopting short-term protections include allowing communities to stockpile resources for feasts or close areas for cultural reasons. Short-term/periodic reserves also may function as partial insurance by enhancing overall ecosystem resilience against catastrophes. Reopened reserves can be protected by management controls that limit the harvest to less than the increase achieved during closure, although at greatly reduced recovery rates.

Resilience

Some habitats and species are better prepared environmental changes or extremes. These include coral communities that handle high sea surface temperature (SST); areas with variable SSTs and carbonate chemistry and areas adjacent to undeveloped low-lying inland areas that coastal habitats can expand into as sea levels rise. Such areas constitute climate change refugia and can potentially better protect biodiversity than more fragile areas. They may also provide fishery benefits, since habitat loss from climate change is a major fishery threat.

Local Practices

Local practices such as overfishing, blast fishing, trawling, coastal development and pollution threaten many marine habitats. These threats decrease ecosystem health and productivity and adversely affect focal and other species. Such practices can also decrease resilience. Some practices that originate beyond reserve boundaries (e.g., runoff) can be mitigated by considering their impacts within broader management frameworks. Areas that are not threatened by such practices and that are adjacent to other unthreatening areas may be better choices for reserves.

Networks

Networks of marine reserves can support both fisheries management and biodiversity conservation. The size, spacing and location of reserves within a network must respect larval dispersal and movement patterns of species that are targeted for protection.

Design

Existing ecological guidelines for designing networks independently focus on achieving either fisheries, biodiversity or climate change objectives or combinations of fisheries and biodiversity or biodiversity and climate change. These three goals have different implications for network design. The most important are reserve size and protection duration (permanent, long term, short term, or periodic closures).

Diversity

Maintaining diversity involves protecting all species. Generally this involves protecting adequate examples of each major habitat (e.g., each type of coral reef, mangrove and seagrass community). Resiliency to threats improves when multiple examples of each habitat are protected.

To address biodiversity or climate change, reserves 4–20 km across are recommended, because they protect larger populations of more species.

Climate Change

Protecting areas that have already proven resilient to ecological changes and/or are relatively well-protected by other protocols are likely to better survive climate change as well.

Fisheries

Reserves 0.5–1 km across export more adults and larvae to fished areas, potentially increasing recruitment and stock replenishment there. Such small reserves are common in the Coral Triangle, where they benefited some fisheries.

Connectivity

Connectivity is the linking of local populations through the (voluntary) dispersal of individuals. Connected reserves are close enough to each other that larvae, juveniles or adults can cross from one to another as their behavior patterns dictate. Connectivity is a key factor in network design, since it allows a disturbed reserve to recover by recruiting individuals from other, potentially over-populated, reserves. Effective networks spaced reserves at distances of <15 km from each other, with smaller reserves spaced more closely.

Most coastal fish species have a bipartite life cycle where larvae are pelagic before settling out of the plankton to live on a reef. While these fish travel varying distances during their life history, their larvae have the potential to move 10s–100s of km, more than the more sedentary adults and juveniles, which have home ranges of <1 m to a few km. Adults and juveniles of some species travel 10s–100s of kilometers as they mature to reach appropriate habitats (e.g., such as coral reef, mangrove and seagrass habitats) or to migrate to spawning areas. When adults and juveniles leave a marine reserve, they become vulnerable to fishing. However, larvae can generally leave a reserve without elevated risk because of their small size and limited fishery exposure. Effective networks account for the movement patterns of target species at each life cycle stage.

Given a strong, consistent current, siting marine reserves upstream increases downstream populations.

Other Protected Marine Categories

Marine reserves are distinct marine parks, but there is some overlap in usage.

Open Ocean Reserves

As of April 2008 no high seas marine reserves had been established. Greenpeace is campaigning for the "doughnut holes" of the western pacific to be declared as marine reserves and for 40 percent of the world's oceans to be so protected.

By country

Australia

- Great Barrier Reef Marine Park is the largest marine park, at 350,000 km² or 217,000 miles² and is partially a marine reserve.

- Australian Whale Sanctuary

- Great Australian Bight Marine Park

- Shark Bay Marine Park

New Zealand

New Zealand has 37 marine reserves spread around the North and South Islands and other outlying islands. These are 'no take' areas where all forms of exploitation are prohibited. Marine reserves are administered by the Department of Conservation. New Zealand's marine environment is more than 15 times larger than its terrestrial area, however only 9.5% of New Zealand's territorial waters are in marine reserves, while 0.4% of New Zealand's exclusive economic zone (EEZ) is currently protected.

British Indian Ocean Territory

- Chagos Marine Reserve is the largest 'no take' marine reserve with more than 540,000 km²

    o It is said to be one of the world's richest marine ecosystems.

    o Hosts the world's biggest living coral structure - the Great Chagos Bank.

    o Home to more than 220 coral species, which is almost half the recorded species of the entire Indian Ocean.

    o Contains more than 1,000 species of reef fish.

National Park

A national park is a park in use for conservation purposes. Often it is a reserve of natural, semi-natural, or developed land that a sovereign state declares or owns. Although individual nations designate their own national parks differently, there is a common idea: the conservation of 'wild nature' for posterity and as a symbol of national pride. An international organization, the International Union for Conservation of Nature (IUCN), and its World Commission on Protected Areas, has defined "National Park" as its *Category II* type of protected areas.

An elephant safari through the Jaldapara National Park in West Bengal, India

While this type of national park had been proposed previously, the United States established the first "public park or pleasuring-ground for the benefit and enjoyment of the people", Yellowstone National Park, in 1872. Although Yellowstone was not officially termed a "national park" in its establishing law, it was always termed such in practice and is widely held to be the first and oldest national park in the world. The first area to use "national park" in its creation legislation was the US's Mackinac Island, in 1875. Australia's Royal National Park, established in 1879, was the world's third official national park. In 1895 ownership of Mackinac Island was transferred to the State of Michigan as a state park and national park status was consequently lost. As a result, Australia's Royal National Park is by some considerations the second oldest national park now in existence.

The largest national park in the world meeting the IUCN definition is the Northeast Greenland National Park, which was established in 1974. According to the IUCN, 6,555 national parks worldwide met its criteria in 2006. IUCN is still discussing the parameters of defining a national park.

National parks are almost always open to visitors. Most national parks provide outdoor recreation and camping opportunities as well as classes designed to educate the public on the importance of conservation and the natural wonders of the land in which the national park is located.

Definitions

Manuel Antonio National Park in Costa Rica was listed by *Forbes* as one of the world's 12 most beautiful national parks.

In 1969, the IUCN declared a national park to be a relatively large area with the following defining characteristics:

- One or several ecosystems not materially altered by human exploitation and occupation, where plant and animal species, geomorphological sites and habitats are of special scientific, educational, and recreational interest or which contain a natural landscape of great beauty;

- Highest competent authority of the country has taken steps to prevent or eliminate exploitation or occupation as soon as possible in the whole area and to effectively enforce the respect of ecological, geomorphological, or aesthetic features which have led to its establishment; and

- Visitors are allowed to enter, under special conditions, for inspirational, educative, cultural, and recreative purposes.

In 1971, these criteria were further expanded upon leading to more clear and defined benchmarks to evaluate a national park. These include:

- Minimum size of 1,000 hectares within zones in which protection of nature takes precedence

- Statutory legal protection

- Budget and staff sufficient to provide sufficient effective protection

- Prohibition of exploitation of natural resources (including the development of dams) qualified by such activities as sport, fishing, the need for management, facilities, etc.

While the term national park is now defined by the IUCN, many protected areas in many countries are called national park even when they correspond to other categories of the IUCN Protected Area Management Definition, for example:

- Swiss National Park, Switzerland: IUCN Ia - Strict Nature Reserve

- Everglades National Park, United States: IUCN Ib - Wilderness Area

- Białowieża National Park, Poland: IUCN II National Park

- Victoria Falls National Park, Zimbabwe: IUCN III - National Monument

- Vitosha National Park, Bulgaria: IUCN IV - Habitat Management Area

- New Forest National Park, United Kingdom: IUCN V - Protected Landscape

- Etniko Ygrotopiko Parko Delta Evrou, Greece: IUCN VI - Managed Resource Protected Area

While national parks are generally understood to be administered by national governments (hence the name), in Australia national parks are run by state governments and predate the Federation of Australia; similarly, national parks in the Netherlands are administered by the provinces. In many countries, including Indonesia, the Netherlands, and the United Kingdom, national parks do not adhere to the IUCN definition, while some areas which adhere to the IUCN definition are not designated as national parks.

History

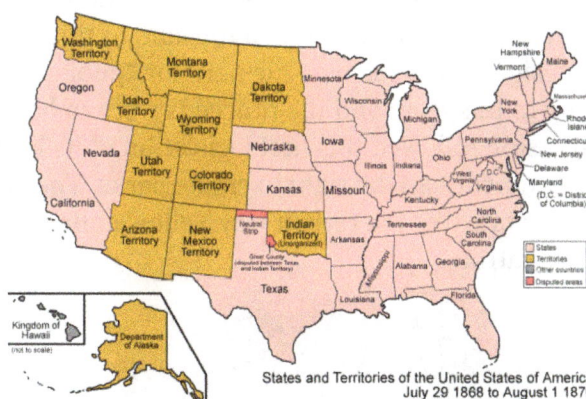

The United States in 1872. When Yellowstone was established, Wyoming, Montana and Idaho were territories, not states. For this reason, the federal government had to assume responsibility for the land, hence the creation of the *national* park.

In 1810, the English poet William Wordsworth described the Lake District as a sort of national property, in which every man has a right and interest who has an eye to perceive and a heart to enjoy.

The painter George Catlin, in his travels through the American West, wrote during the 1830s that the Native Americans in the United States might be preserved

(by some great protecting policy of government) ...in a *magnificent park* ...A *nation's Park*, containing man and beast, in all the wild and freshness of their nature's beauty!

The first effort by the Federal government to set aside such protected lands was on April 20, 1832, when President Andrew Jackson signed legislation that the 22nd United States Congress had enacted to set aside four sections of land around what is now Hot Springs, Arkansas, to protect the natural, thermal springs and adjoining mountainsides for the future disposal of the U.S. government. It was known as Hot Springs Reservation, but no legal authority was established. Federal control of the area was not clearly established until 1877.

Yosemite Valley, Yosemite National Park, in California

John Muir is today referred to as the "Father of the National Parks" due to his work in Yosemite. He published two influential articles in The Century Magazine, which formed the base for the subsequent legislation.

President Abraham Lincoln signed an Act of Congress on July 1, 1864, ceding the Yosemite Valley and the Mariposa Grove of Giant Sequoias (later becoming Yosemite National Park) to the state of California. According to this bill, private ownership of the land in this area was no longer possible. The state of California was designated to manage the park for "public use, resort, and recreation". Leases were permitted for up to ten years and the proceeds were to be used for conservation and improvement. A public discussion followed this first legislation of its kind and there was a heated debate over whether the government had the right to create parks. The perceived mismanagement of Yosemite by the Californian state was the reason why Yellowstone at its establishment six years later was put under national control.

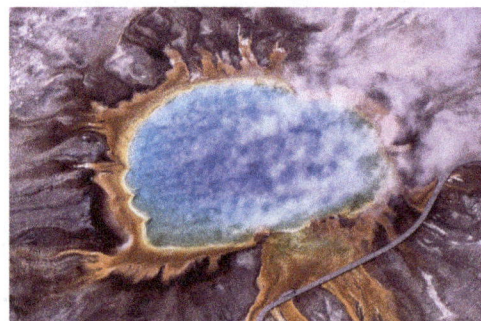

Grand Prismatic Spring in Yellowstone National Park, Wyoming; Yellowstone was the first national park in the world.

In 1872, Yellowstone National Park was established as the United States' first national park, being also the world's first national park. In some European countries, however, national protection and nature reserves already existed, such as Drachenfels (Germany, 1822) and a part of Forest of Fontainebleau (France, 1861).

Yellowstone was part of a federally governed territory. With no state government that could assume stewardship of the land so the federal government took on direct responsibility for the park, the official first national park of the United States. The combined effort and interest of conservationists, politicians and the Northern Pacific Railroad ensured the passage of enabling legislation by the United States Congress to create Yellowstone National Park. Theodore Roosevelt, already an active campaigner and so influential, as good stump speakers were highly necessary in the pre-telecommunications era, was highly influential in convincing fellow Republicans and big business to back the bill.

American Pulitzer Prize-winning author Wallace Stegner wrote:

National parks are the best idea we ever had. Absolutely American, absolutely democratic, they reflect us at our best rather than our worst.

In his book *Dispossessing the Wilderness: Indian Removal and the Making of the National Parks*, Mark David Spence made the point that in order to create these uninhabited spaces, the United States first had to disposess the Indians who were living in them.

Even with the creation of Yellowstone, Yosemite, and nearly 37 other national parks and monuments, another 44 years passed before an agency was created in the United States to administer these units in a comprehensive way – the U.S. National Park Service (NPS). The 64th United States Congress passed the National Park Service Organic Act, which President Woodrow Wilson signed into law on August 25, 1916. Of the 413 sites managed by the National Park Service of the United States, only 59 carry the designation of National Park.

Following the idea established in Yellowstone, there soon followed parks in other nations. In Australia, the Royal National Park was established just south of Sydney on April 26, 1879, becoming the world's second official national park (actually the 3rd: Mackinac National Park in Michigan was created in 1875 as a national park but was later transferred to the state's authority in 1895, thus losing its official "national park" status). Rocky Mountain National Park became Canada's first national park in 1885. Argentina became the third country in the Americas to create a national park system, with the creation of the Nahuel Huapi National Park in 1934, through the initiative of Francisco Moreno. New Zealand established Tongariro National Park in 1887. In Europe, the first national parks were a set of nine parks in Sweden in 1909, followed by the Swiss National Park in 1914. Europe has some 359 national parks as of 2010. Africa's first national park was established in 1925 when Albert I of Belgium designated an area of what is now Democratic Republic of Congo centred on the Virunga Mountains as the Albert National Park (since renamed Virunga National Park). In 1973, Mount Kilimanjaro was classified as a National Park and was opened to public access in 1977. In 1926, the government of South Africa designated Kruger National Park as the nation's first national park, although it was an expansion of the earlier Sabie Game Reserve established in 1898 by President Paul Kruger of the old

South African Republic, after whom the park was named. After World War II, national parks were founded all over the world. The Vanoise National Park in the Alps was the first French national park, created in 1963 after public mobilization against a touristic project.

The world's first national park service was established May 19, 1911, in Canada. The Dominion Forest Reserves and Parks Act placed the dominion parks under the administration of the Dominion Park Branch (now Parks Canada). The branch was established to "protect sites of natural wonder" to provide a recreational experience, centered on the idea of the natural world providing rest and spiritual renewal from the urban setting. Canada now has the largest protected area in the world with 377,000 km² of national park space. In 1989, the Qomolangma National Nature Preserve (QNNP) was created to protect 3.381 million hectares on the north slope of Mount Everest in the Tibet Autonomous Region of China. This national park is the first major global park to have no separate warden and protection staff—all of its management being done through existing local authorities, allowing a lower cost basis and a larger geographical coverage (in 1989 when created, it was the largest protected area in Asia). It includes four of the six highest mountains Everest, Lhotse, Makalu, and Cho Oyu. The QNNP is contiguous to four Nepali national parks, creating a transborder conservation area equal in size to Switzerland.

- National parks

Economic Ramifications

Countries with a large nature-based tourism industry, such as Costa Rica, often experience a huge economic effect on park management as well as the economy of the country as a whole.

Tourism

Tourism to national parks has increased considerably over time. In Costa Rica for example, a megadiverse country, tourism to parks has increased by 400% from 1985 to 1999. The term *national park* is perceived as a brand name that is associated with nature-based tourism and it symbolizes "high quality natural environment and well-design tourism infrastructure".

Staff

The duties of a park ranger are to supervise, manage, and/or perform work in the conservation and use of Federal park resources. This involves functions such as park conservation; natural, historical, and cultural resource management; and the development and operation of interpretive and recreational programs for the benefit of the visiting public. Park rangers also have fire fighting responsibilities and execute search and rescue missions. Activities also include heritage interpretation to disseminate information to visitors of general, historical, or scientific information. Management of resources such as wildlife, lakeshores, seashores, forests, historic buildings, battlefields, archeological properties, and recreation areas are also part of the job of a park ranger. Since the establishment of the National Park Service in the US in 1916, the role of the park ranger has shifted from merely being a custodian of natural resources to include several activities that are associated with law enforcement. They control traffic and investigate violations, complaints, trespass/encroachment, and accidents.

Geopark

World locator map for Geoparks included in the UNESCO Global Geoparks Network—GGN.

A Geopark is a unified area that advances the protection and use of geological heritage in a sustainable way, and promotes the economic well-being of the people who live there. There are Global Geoparks and National Geoparks.

Concepts

A *Global Geopark* is a unified area with geological heritage of international significance. Geoparks use that heritage to promote awareness of key issues facing society in the context of the dynamic planet we all live on. Many geoparks promote awareness of geological hazards, including volcanoes, earthquakes and tsunamis and many help prepare disaster mitigation strategies among local communities. Geoparks hold records of past climate change and are educators on current climate change as well as adopting a best practise approach to utilising renewable energy and employing the best standards of "green tourism".Tourism industry promotion in geopark, as a geographically sustainable and applicable tourism model, aims to sustain, or even enhance, the geographical character of a place.

Geoparks also inform about the sustainable use and need for natural resources, whether they are mined, quarried or harnessed from the surrounding environment while at the same time promoting respect for the environment and the integrity of the landscape. Geoparks are not a legislative designation though the key heritage sites within a geopark are often protected under local, regional or national legislation as appropriate. The multidisciplinary nature of the concept of geopark and tourism promotion in geoparks differentiates itself from other models of sustainable tourism. In fact, sustainable tourism promotion within geopark actually encompasses many of the subdivisions of sustainable tourism including: geo-tourism (geo-site tourism: as a basic factor), community-based tourism and integrated rural tourism (as a vital needs), ecotourism, cultural heritage tourism, etc.

Global Network and UNESCO

The Global Geoparks Network (GGN) is supported by United Nations Educational, Scientific and Cultural Organization (UNESCO). Many national geoparks and other local geoparks projects also exist which are not included in the Global Geoparks Network.

The geoparks initiative was launched by UNESCO in response to the perceived need for an international initiative that recognizes sites representing an earth science interest. Global Geoparks Network aims at enhancing the value of such sites while at the same time creating employment and promoting regional economic development. The Global Geoparks Network works in synergy with UNESCO's World Heritage Centre and Man and the Biosphere (MAB) World Network of Biosphere Reserves.

Qualification

The Global Geoparks Network (GGN) is a UNESCO activity established in 1998. According to UNESCO, for a geopark to apply to be included in the GGN, it needs to:

- have a management plan designed to foster socio-economic development that is sustainable based on geotourism

- demonstrate methods for conserving and enhancing geological heritage and provide means for teaching geoscientific disciplines and broader environmental issues

- have joint proposals submitted by public authorities, local communities and private interests acting together, which demonstrate the best practices with respect to Earth heritage conservation and its integration into sustainable development strategies.

Members of The GGN

Nature Park

A nature park or natural park is a landscape protected by means of long-term planning, use and agriculture. These valuable landscapes are preserved in their present state and promoted for tourism purposes.

In most countries nature parks are subject to legally regulated protection, which is part of their conservation laws.

The South Mountain Reservation in the New York Metropolitan Area

A nature park is *not* the same as a national park, which is defined by the International Union for Conservation of Nature (IUCN) and its World Commission on Protected Areas as a category II type of protected area. Nor is it the same as a nature reserve which may fall into different IUCN categories depending on the level of legal protection afforded, but which is more strictly protected than a nature park. However some nature parks have later been turned into national parks.

International Nature Parks

The first international nature park in Europe, the present-day Pieniny National Park was founded jointly by Poland and Slovakia in 1932.

- *European Nature Parks*: Cross-border plans and projects are carried out under the *Europarc* umbrella.

- *Protected Area Network of Parks* (PANPark), certification by the WWF initiated network which are aimed at combining the preservation of wilderness with tourism

Europe

Austria

There are currently 47 nature parks in Austria with a total area of around 500,000 ha (as at April 2010). They are host to nearly 20 million visitors annually. The designation of "nature park" is awarded by the respective state governments. To achieve this award, the 4 pillars of a nature park have to be met: conservation, recreation, education and regional development.

Overview map of Austria's nature parks

Association of Austrian Nature Parks (VNÖ)

In 1995 all the Austrian nature parks agreed to be represented by the Association of Austrian Nature Parks (*Verband der Naturparke Österreichs*) or VNÖ.

Currently there are nature parks in the following states:

- Burgenland:
  - Geschriebenstein-Irottkö
  - Landseer Berge
  - Raab-Örseg-Goricko
  - Weinidylle
  - Neusiedler See-Leithagebirge
  - Rosalia-Kogelberg
- Carinthia
  - Dobratsch
  - Weißensee
- Lower Austria:
  - Blockheide-Eibenstein-Gmünd
  - Buchenberg-Waidhofen an der Ybbs
  - Dobersberg-Thayatal
  - Eichenhain bei Klosterneuburg
  - Eisenwurzen NÖ
  - Falkenstein
  - Föhrenberge-Mödling
  - Geras
  - Heidenreichsteiner Moor
  - Hochmoor Schrems
  - Hohe Wand Nature Park
  - Jauerling-Wachau
  - Kamptal-Schönberg
  - Leiser Berge
  - Mannersdorf-Wüste
  - Nordwald-Bad Großpertholz

-
  - Ötscher-Tormäuer
  - Purkersdorf-Sandstein-Wienerwald
  - Seebenstein-Türkensturz
  - Sierningtal-Flatzer Wand
  - Sparbach (ältester Naturpark, seit 1962)
- Upper Austria:
  - Mühlviertel
  - Obst-Hügel-Land
- Salzburg
  - Buchberg
  - Riedingtal in Zederhaus
  - Weißbach bei Lofer
- Styria:
  - Almenland
  - Mürzer Oberland
  - Sölktäler
  - Steirische Eisenwurzen
  - Pöllauer Tal
  - Südsteirisches Weinland
  - Zirbitzkogel-Grebenzen
- Tyrol
  - Alpenpark Karwendel
  - Kaunergrat (Pitztal-Kaunertal)
  - Tyrolean Lechtal
  - Zillertal Alps
  - Ötztal

Belgium

In Belgium, there are two different structures. In Flanders, their name is Regionale Landschappen and in Wallonia, the Natural Parks. There are 17 Regionale Landschappen in Flanders and 9 Natural Parks in Wallonia.

- Fédération des Parcs naturels de Wallonie
- Regionale Landschappen

Croatia

In Croatia there is a total of eight national parks and eleven nature parks. Under nature park protection are the following regions:

- Biokovo
- Kopački rit
- Lonjsko polje
- Medvednica
- Papuk
- Telašćica
- Učka
- Velebit
- Vransko jezero
- Žumberak-Samoborsko gorje
- Lastovo

Czech Republic

In the Czech Republic a Nature Park (Přírodní Park)is defined as a large area serving the protection of a landscape against activities that could decrease its natural and esthetic value. They can be established by any State Environment Protection body.

France

Germany

The *Nature park* is one of the options for area-based nature conservation provided for under the Federal Nature Conservation Act (the *BNatSchG*). On 6 June 1956 in the former capital city of Bonn at the annual meeting of the Nature Reserve Association (in the presence of President Theodor Heuss and Minister Heinrich Lübke)., the environmentalist and entrepreneur, Alfred Toepfer, presented a programme developed jointly with the Central Office for Nature Conservation and Landscape Management and other institutions to set up (initially) 25 nature parks in West Germany. Five percent of the area of the old Federal Republic of Germany was to be spared from major environmental damage as a result.

Nature parks in Germany

Definition of Nature Parks in Germany

The definition of the category of nature park is laid down in federal law (§ 27 of the BNatSchG). Details, especially with regard to the identification, investigation or recognition as a nature park vary in each state depending on the provisions of local conservation law.

§ 27 of the BnatSchG determines that natural parks are large areas that are to be developed and managed as a single unit, that consist mainly of protected landscapes or nature reserves, that have a large variety of species and habitats and that have a landscape that exhibits a variety of uses.

In nature parks, the aim is to strive for environmentally sustainable land use and they should be especially suitable for recreation and for sustainable tourism because of their topographical features.

The underlying idea is a "protection through usage", so the acceptance and participation of the population in the protection of the cultural landscape and nature is very important. In doing so the nature conservation and the needs of recreation users should be linked so that both sides benefit: sustainable tourism with respect for the value of nature and landscape is paramount.

Basically all actions, interventions and projects that would be contrary to the purpose of conservation are prohibited.

Nature parks are to be considered in zoning and must be represented and considered in local development plans. This is called an acquisition memorandum. They are binding and cannot be waived because of a higher common good.

The sponsors of nature parks are usually clubs or local special purpose associations.

The German nature parks come together in the Association of German Nature Parks.

In Germany today there are 101 nature parks (as at: March 2009), that occupy some 25% of the land area. They are an important building block for nature conservation and help to preserve the sites of natural beauty, cultural landscapes, rare species and biotopes and to make them accessible to later generations.

Italy

South Tyrol has 8 nature parks and part of a national park

- Schlern-Rosengarten
- Texelgruppe
- Puez-Geisler
- Fanes-Sennes-Prags
- Trudner Horn
- Sexten Dolomites
- Rieserferner-Ahrn
- Sarntal Alps
- Stelvio

Hungary

- Geschriebenstein-Írottkő (cross-border park with Burgenland in Austria)
- Raab-Örseg-Goricko (cross-border park with Burgenland in Austria and Goričko in Slovenia)

Switzerland

A view in the Swiss National Park.

In Switzerland the establishment of regional nature parks is regulated by the Federal Act on the Protection of Nature and Cultural Heritage. The three categories are:

- National parks (the Swiss National Park)
- Regional nature parks (sixteen parks)

- o   Aargau Jura Park

- o   Parc Ela

- o   Thal Nature Park

- o   Entlebuch Biosphere

- o   Val Müstair Biosphere

- o   Etc.

- Nature experience parks (the Wildnispark Zurich Sihlwald)

National parks and nature experience parks have very strict protected areas, something which does not exist in regional nature parks. The latter focus much more on striking a balance in the level of support between nature conservation and the regional economy.

References

- Schrijver, Nico (2010). Development Without Destruction: The UN and Global Resource Management. United Nations Intellectual History Project Series. Bloomington, IN: Indiana University Press. p. 116. ISBN 978-0-253-22197-1.

- Strong, Maurice; Introduction by Kofi Annan (2001). Where on Earth are We Going? (Reprint ed.). New York, London: Texere. pp. 120–136. ISBN 1-58799-092-X.

- Turner, D.B. (1994). Workbook of atmospheric dispersion estimates: an introduction to dispersion modeling (2nd ed.). CRC Press. ISBN 1-56670-023-X. www.crcpress.com

- Beychok, M.R. (2005). Fundamentals Of Stack Gas Dispersion (4th ed.). author-published. ISBN 0-9644588-0-2. Air-dispersion.com

- European Environment Agency Protected areas in Europe – an overview In: EEA Report No 5/2012 Kopenhagen: 2012 ISBN 978-92-9213-329-0 ISSN 1725-9177 pdf doi=10.2800/55955

- McMillan, A.J.S.; Horobin, J.F. (1995), Christmas Cacti : The genus Schlumbergera and its hybrids (p/b ed.), Sherbourne, Dorset: David Hunt, ISBN 978-0-9517234-6-3

- "Papahānaumokuākea (Expansion) Marine National Monument, United States". MPAtlas. Marine Conservation Institute. Retrieved 2016-09-02.

- Harary, David. "Joint-Value Creation Between Marine Protected Areas and the Private Sector" (PDF). thinkcds.org. Center for Development and Strategy. Retrieved 26 July 2016.

- "Protection of the South Orkney Islands southern shelf". https://www.ccamlr.org. Commission for the Conservation of Antarctic Marine Living Resources. Retrieved 29 October 2016. External link in |website= (help)

- "CCAMLR to create world's largest Marine Protected Area". Commission for the Conservation of Antarctic Marine Living Resources. 29 October 2016. Retrieved 29 October 2016.

- "Conservationists call for UK to create world's largest marine reserve". The Guardian. 15 February 2015. Retrieved 4 January 2016.

- Weiland, Paul S. (Spring 1997). "Amending the National Environmental Policy Act: Federal Environmental Protection in the Twenty-First Century" (PDF). Journal of Land Use & Environmental Law: 275–301. Retrieved 21 February 2015.

- Adler, Jonathan H. (2015-12-15). "GAO hits EPA for 'covert propaganda' to promote 'waters of the United States' (WOTUS) rule". The Washington Post. ISSN 0190-8286. Retrieved 2015-12-16.

- *Schlanger, Zoë (August 7, 2015). "EPA Causes Massive Spill of Mining Waste Water in Colorado, Turns Animas River Bright Orange". Retrieved August 10, 2015.*

- *Kolb, Joseph J. (August 10, 2015). "'They're not going to get away with this': Anger mounts at EPA over mining spill". Retrieved August 10, 2015.*

- *"EPA's Budget and Spending | Planning Budget Results". US EPA. United States Environmental Protection Agency. 2014-03-04. Retrieved 2014-03-28.*

- *"Lost in Yonkers?Head South from Hillview" (PDF). Weekly Pipeline. NYC Department of Environmental Protection. II (80). 2011-07-11. Retrieved 2014-10-30.*

- *"Federal government may scrap much-derided $1.6 billion reservoir cap". NY Daily News. 2011-08-21. Retrieved 2014-10-30.*

- *"EPA Library Network Strategic Plan FY 2012– FY 2014 | EPA National Library Network | US EPA". Epa. gov. 2006-06-28. Retrieved 2012-10-21.*

6

Organizations Promoting Environmental Protection

Conserving the environment is a global concern and authorities and organizations across the globe are working towards it. This chapter incorporates topics like United Nations environment programme, United Nations conference on the human environment, ministry of environmental protection of the people's republic of China, United States environmental protection agency and conservation international, etc. to shed light on organizations promoting environmental protection.

United Nations Environment Programme

The United Nations Environment Programme (UNEP) is an agency of United Nations and coordinates its environmental activities, assisting developing countries in implementing environmentally sound policies and practices. It was founded by Maurice Strong, its first director, as a result of the United Nations Conference on the Human Environment (Stockholm Conference) in June 1972 and has its headquarters in the Gigiri neighborhood of Nairobi, Kenya. UNEP also has six regional offices and various country offices.

Its activities cover a wide range of issues regarding the atmosphere, marine and terrestrial ecosystems, environmental governance and green economy. It has played a significant role in developing international environmental conventions, promoting environmental science and information and illustrating the way those can be implemented in conjunction with policy, working on the development and implementation of policy with national governments, regional institutions in conjunction with environmental non-governmental organizations (NGOs). UNEP has also been active in funding and implementing environment related development projects.

UNEP has aided in the formulation of guidelines and treaties on issues such as the international trade in potentially harmful chemicals, transboundary air pollution, and contamination of international waterways.

The World Meteorological Organization and UNEP established the Intergovernmental Panel on Climate Change (IPCC) in 1988. UNEP is also one of several Implementing Agencies for the Global Environment Facility (GEF) and the Multilateral Fund for the Implementation of the Montreal Protocol, and it is also a member of the United Nations Development Group. The International Cyanide Management Code, a program of best practice for the chemical's use at gold mining operations, was developed under UNEP's aegis.

Executive Director

UNEP's current Executive Director Erik Solheim succeeded the previous director Achim Steiner in 2016.

On 15 March 2006, the former Secretary-General of the United Nations, Kofi Annan, nominated Achim Steiner, former Director General of the IUCN to the position of Executive Director. The UN General Assembly followed Annan's proposal and elected him.

The position was held for 17 years (1975–1992) by Dr. Mostafa Kamal Tolba, who was instrumental in bringing environmental considerations to the forefront of global thinking and action. Under his leadership, UNEP's most widely acclaimed success—the historic 1987 agreement to protect the ozone layer—the Montreal Protocol was negotiated.

During December 1972, the UN General Assembly unanimously elected Maurice Strong to head UNEP. Also Secretary General of both the 1972 United Nations Conference on the Human Environment, which launched the world environment movement, and the 1992 Earth Summit, Strong has played a critical role is globalizing the environmental movement.

List of Executive Directors

#	Picture	Name (Birth–Death)'	Nationality	Took office	Left office
6		**Erik Solheim** (born 1955)	Norway	2016	**Present**
5		**Achim Steiner** (born 1961)	Germany	2006	2016
4		**Klaus Töpfer** (born 1938)	Germany	1998	2006

3		Elizabeth Dowdeswell (born 1944)	Canada	1992	1998
2		Mostafa Kamal Tolba (1922-2016)	Egypt	1975	1992
1		Maurice Strong (1929-2015)	Canada	1972	1975

Structure

UNEP's structure includes seven substantive Divisions:

- Early Warning and Assessment (DEWA)
- Environmental Policy Implementation (DEPI)
- Technology, Industry and Economics (DTIE)
- Regional Cooperation (DRC)
- Environmental Law and Conventions (DELC)
- Communications and Public Information (DCPI)
- Global Environment Facility Coordination (DGEF)

Internationtal Years

The year 2007 was declared (International) Year of the Dolphin by the United Nations and UNEP.

(International) Patron of the Year of the Dolphin was H.S.H. Prince Albert II of Monaco, with Special Ambassador to the cause being Nick Carter, of the Backstreet Boys.

2010 was designated the International Year of Biodiversity and presented an opportunity to enhance knowledge of ecosystems and their services.

In 2011 the UN celebrated the International Year of Forests.

In 2012, the International Year for Sustainable Energy for All.

2013 has been designated as the International Year of Water Cooperation.

(See international observance and list of environmental dates.)

Reports

UNEP publishes many reports, atlases and newsletters. For instance, the fifth Global Environment Outlook (GEO-5) assessment is a comprehensive report on environment, development and human well-being, providing analysis and information for policy makers and the concerned public. One of many points in the GEO-5 warns that we are living far beyond our means. It notes that the human population is now so large that the amount of resources needed to sustain it exceeds what is available.

In June 2010, a report from UNEP declared that a global shift towards a vegan diet was needed to save the world from hunger, fuel shortages and climate change.

Reform

Following the publication of Fourth Assessment Report of the Intergovernmental Panel on Climate Change (IPCC) in February 2007, a "Paris Call for Action" read out by French President Jacques Chirac and supported by 46 countries, called for the United Nations Environment Programme to be replaced by a new and more powerful "United Nations Environment Organization (UNEO)", also called Global Environment Organisation now supported by French President Nicolas Sarkozy and German Chancellor Angela Merkel, to be modelled on the World Health Organization. The 46 countries included the European Union nations, but notably did not include the United States, Saudi Arabia, Russia, and China, the top four emitters of greenhouse gases.

In December 2012, following the Rio+20 Summit, a decision by the General Assembly of the United Nations to 'strengthen and upgrade' the UN Environment Programme (UNEP) and establish universal membership of its governing body was confirmed.

Main Activities

UNEP's main activities are related to:

- climate change;
    - including the Territorial Approach to Climate Change (TACC);
- disasters and conflicts;
- ecosystem management;
- environmental governance;
- environment under review;
- harmful substances; and
- resource efficiency.

Notable World Projects

UNEP has sponsored the development of solar loan programs, with attractive return rates, to buffer the initial deployment costs and entice consumers to consider and purchase solar PV systems. The most famous example is the solar loan program sponsored by UNEP helped 100,000 people finance solar power systems in India. Success in India's solar program has led to similar projects in other parts of the developing world like Tunisia, Morocco, Indonesia and Mexico.

UNEP sponsors the Marshlands project in the Middle East . In 2001, UNEP alerted the international community to the destruction of the Marshlands when it released satellite images showing that 90 percent of the Marshlands had already been lost. The UNEP "support for Environmental Management of the Iraqi Marshland" commenced in August 2004, in order to manage the Marshland area in an environmentally sound manner.

In order to ensure full participation of global communities, UNEP works in an inclusive fashion that brings on board different societal cohorts. UNEP has a vibrant programme for young people known as Tunza. Within this program are other projects like the AEO for Youth.

Glaciers Shrinking

Glaciers are shrinking at record rates and many could disappear within decades, the U.N. Environment Programme said on March 16, 2008. The scientists measuring the health of almost 30 glaciers around the world found that ice loss reached record levels in 2006. On average, the glaciers shrank by 4.9 feet in 2006, the most recent year for which data are available. The most severe loss was recorded at Norway's Breidalblikkbrea glacier, which shrank 10.2 feet in 2006. Glaciers lost an average of about a foot of ice a year between 1980 and 1999. But since the turn of the millennium the average loss has increased to about 20 inches.

Electric Vehicles

At the fifth Magdeburg Environmental Forum held from 3–4 July 2008, in Magdeburg, Germany, UNEP and car manufacturer Daimler called for the establishment of infrastructure for electric vehicles. At this international conference, 250 high-ranking representatives from ce, politics and non-government organizations discussed solutions for future road transportation under the motto of "Sustainable Mobility–the Post-2012 CO2 Agenda".

United Nations Conference on the Human Environment

The United Nations Conference on the Human Environment was held in Stockholm, Sweden from June 5–16 in 1972.

When the UN General Assembly decided to convene the 1972 Stockholm Conference, at the initiative of the Government of Sweden to host it, UN Secretary-General U Thant invited Maurice Strong to lead it as Secretary-General of the Conference, as the Canadian diplomat (under Pierre Trudeau) had initiated and already worked for over two years on the project.

Contents

History

Sweden first suggested to ECOSOC in 1968 the idea of having a UN conference to focus on human interactions with the environment. ECOSOC passed resolution 1346 supporting the idea. General

Assembly Resolution 2398 in 1969 decided to convene a conference in 1972 and mandated a set of reports from the UN secretary-general suggesting that the conference focus on "stimulating and providing guidelines for action by national government and international organizations" facing environmental issues.

Stockholm Declaration

The meeting agreed upon a Declaration containing 26 principles concerning the environment and development; an Action Plan with 109 recommendations, and a Resolution. Principles of the Stockholm Declaration:

1. Human rights must be asserted, apartheid and colonialism condemned

2. Natural resources must be safeguarded

3. The Earth's capacity to produce renewable resources must be maintained

4. Wildlife must be safeguarded

5. Non-renewable resources must be shared and not exhausted

6. Pollution must not exceed the environment's capacity to clean itself

7. Damaging oceanic pollution must be prevented

8. Development is needed to improve the environment

9. Developing countries therefore need assistance

10. Developing countries need reasonable prices for exports to carry out environmental management

11. Environment policy must not hamper development

12. Developing countries need money to develop environmental safeguards

13. Integrated development planning is needed

14. Rational planning should resolve conflicts between environment and development

15. Human settlements must be planned to eliminate environmental problems

16. Governments should plan their own appropriate population policies

17. National institutions must plan development of states' natural resources

18. Science and technology must be used to improve the environment

19. Environmental education is essential

20. Environmental research must be promoted, particularly in developing countries

21. States may exploit their resources as they wish but must not endanger others

22. Compensation is due to states thus endangered

23. Each nation must establish its own standards

24. There must be cooperation on international issues

25. International organizations should help to improve the environment

26. Weapons of mass destruction must be eliminated

One of the seminal issue that emerged from the conference is the recognition for poverty alleviation for protecting the environment. The Indian Prime Minister Indira Gandhi in her seminal speech in the conference brought forward the connection between ecological management and poverty alleviation.

Some argue that this conference, and more importantly the scientific conferences preceding it, had a real impact on the environmental policies of the European Community (that later became the European Union). For example, in 1973, the EU created the Environmental and Consumer Protection Directorate, and composed the first Environmental Action Program. Such increased interest and research collaboration arguably paved the way for further understanding of global warming, which has led to such agreements as the Kyoto Protocol and the Paris Agreement, and has given a foundation of modern environmentalism.

Ministry of Environmental Protection of the People's Republic of China

The Ministry of Environmental Protection of the People's Republic of China (MEP), formerly State Environmental Protection Administration (SEPA), is a cabinet-level ministry in the executive branch of the Government of China. It replaced the SEPA during the March 2008 National People's Congress sessions in Beijing.

The Ministry is the nation's environmental protection department charged with the task of protecting China's air, water, and land from pollution and contamination. Directly under the State Council, it is empowered and required by law to implement environmental policies and enforce environmental laws and regulations. Complementing its regulatory role, it funds and organizes research and development. In addition, it also serves as China's nuclear safety agency.

History

In 1972, Chinese representatives attended the First United Nations Conference on the Human Environment, held in Sweden. The next year, 1973, saw the establishment of the Environmental Protection Leadership Group. In 1983, the Chinese government announced that environmental protection would become a state policy. In 1998, China went through a disastrous year of serious flooding, and the Chinese government upgraded the Leading Group to a ministry-level agency, which then became the State Environmental Protection Administration.

Organization

There are 12 offices and departments under MEP, all at the *si* (司) level in the government ranking system. They carry out regulatory tasks in different areas and make sure that the agency is functioning accordingly:

Department Structure

Department	Chinese Name
General Administrative Office	(办公厅)
Department of Human Resources & Institutional Affairs	(行政体制与人事司)
Department of Planning and Finance	(规划与财务司)
Department of Policies, Laws and Regulations	(政策法规司)
Department of Science & Technology and Standards	(科技标准司)
Pollution Control Office	(污染控制司)
Natural Ecosystem Protection Office	(自然生态保护司)
Department of Environmental Impact Assessment	(环境影响评价管理司)
International Cooperation Office	(国际合作司)
Department of Nuclear Safety	(核安全管理司)
Environmental Inspection Office	(环境监察局)
Office of Agency & Party Affairs	(机关党委)

Leadership

Position	Name
Administrator/ Minister:	Chen Jining
Vice-Minister:	Pan Yue (潘岳)
Head of Discipline:	Zhu Guangyao (祝光耀)
Vice-Minister:	Zhang Lijun (张力军)
Vice-Minister:	Wu Xiaochun (吴晓青)
Vice-Minister:	Zhou Jian (周建)
Vice-Minister, Bureau Chief for Nuclear Safety:	Li Ganjie (李干杰)

Minister Xie Zhenhua resigned in December 2005 amidst an industrial pollution scandal by PetroChina, a Chinese national oil company, on the Songhua River in the northeastern province Heilongjiang; local environmental protection officials were accused of protectionism, while senior officials at SEPA were blamed for their underestimating and ignoring the matter.

The Vice-Minister, Pan Yue (潘岳), who has served in SEPA with Xie and is still in power, has been one of the most vocal high-level officials in the Chinese government critical of the current development model. He warned during an interview with the German newspaper Der Spiegel in 2005 that "the Chinese miracle will end soon" if sustainable issues were not addressed urgently.

Regional Centers

In 2006, SEPA opened five regional centers to help with local inspections and enforcement. Today, the five centers are direct affiliates of MEP:

Region	Head Office	Enforcement Area
Eastern Center	Nanjing	Shanghai, Jiangsu, Zhejiang, Anhui, Fujian, Jiangxi, and Shandong.
Southern Center	Guangzhou	Hunan, Hubei, Guangdong, Guangxi, and Hainan.
Northwestern Center	Xi'an	Shaanxi, Gansu, Qinghai, Xinjiang, and Ningxia.
Southwestern Center	Chengdu	Chongqing, Sichuan, Guizhou, Yunnan, and Tibet.
Northeastern Center	Shenyang	Liaoning, Jining, and Heilongjiang.
MEP headquarters	Beijing	Beijing, Tianjing, Hebei, Henan, Shanxi, and Inner Mongolia.

Areas of Activities

MEP regulates water quality, ambient air quality, solid waste, soil, noise, radioactivity. In the area of R&D activities, MEP has funded a series of "Key Laboratories" in different parts of the country, including: Laboratory for Urban Air Particles Pollution Prevention and Control for Environmental Protection, Laboratory on Environment and Health, Laboratory on Industrial Ecology, Laboratory on Wetland Ecology and Vegetation Recovery, and Laboratory on Biosafety.

In addition, MEP also administers engineering and technical research centers related to environmental protection, including: Center for Non-ferrous Metal Industrial Pollution Control, Center for Clean Coal and Ecological Recovery of Mines, Center for Industrial Waste Water Pollution Control, Center for Industrial Flue Gas Control, Center for Hazardous Waste Treatment, and Center for Solid Waste Treatment and Disposal of Mines.

China is experiencing an increase in environmental complaints: In 2005, there were 51,000 disputes over environmental pollution, according to SEPA minister Zhou Shengxian. From 2001 to 2005, Chinese environmental authorities received more than 2.53 million letters and 430,000 visits by 597,000 petitioners seeking environmental redress.

In the Media

Economic Development

Vice minister Pan Yue, a former journalist, said in an interview with http://www.chinadialogue.net that the fundamental cause of the worsening global environmental crisis "...is the capitalist system.

The environmental crisis has become a new means of transferring the economic crisis.". He believes China's role in the environmental crisis "… has arisen, basically, because our mode of economic modernisation has been copied from western, developed nations. In 30 years, China has achieved economic results that took a century to attain in the West. But we have also concentrated a century's worth of environmental issues into those 30 years. While becoming the world leader in GDP growth and foreign investment, we have also become the world's number one consumer of coal, oil and steel – and the largest producer of CO_2 and chemical oxygen demand (COD) emissions.".

North Korean Nuclear Test

After a North Korean nuclear test, the Ministry of Environmental Protection issued a press release "to reassure residents that no radioactive particles had been detected in aerosol samples collected" in the Northeast region.

List of Ministers

№	Name	Took office	Left office
Director of State Environmental Protection Agency			
1	Qu Geping (zh)	1987	June 1993
2	Xie Zhenhua	June 1993	March 1998
Director of State Environmental Protection Administration			
(2)	Xie Zhenhua	March 1998	December 2005
3	Zhou Shengxian	December 2005	March 2008
Minister of Environmental Protection			
(3)	Zhou Shengxian	March 2008	February 2015
4	Chen Jining	February 2015	Incumbent

United States Environmental Protection Agency

The United States Environmental Protection Agency (EPA or sometimes USEPA) is an agency of the U.S. federal government which was created for the purpose of protecting human health and the environment by writing and enforcing regulations based on laws passed by Congress. The EPA was proposed by President Richard Nixon and began operation on December 2, 1970, after Nixon signed an executive order. The order establishing the EPA was ratified by committee hearings in the House and Senate. The agency is led by its Administrator, who is appointed by the president and approved by Congress. The current administrator is Gina McCarthy. The EPA is not a Cabinet department, but the administrator is normally given cabinet rank.

The EPA has its headquarters in Washington, D.C., regional offices for each of the agency's ten regions, and 27 laboratories. The agency conducts environmental assessment, research, and education. It has the responsibility of maintaining and enforcing national standards under a variety of environmental laws, in consultation with state, tribal, and local governments. It delegates some permitting, monitoring, and enforcement responsibility to U.S. states and the federally recognized tribes. EPA enforcement powers include fines, sanctions, and other measures. The agency also works with industries and all levels of government in a wide variety of voluntary pollution prevention programs and energy conservation efforts.

The agency has approximately 15,193 full-time employees and engages many more people on a contractual basis. More than half of EPA human resources are engineers, scientists, and environmental protection specialists; other groups include legal, public affairs, financial, and information technologists.

History

Stacks emitting smoke from burning discarded automobile batteries, photo taken in Houston in 1972 by Marc St. Gil (cs), official photographer of recently founded EPA

Same smokestacks in 1975 after the plant was closed in a push for greater environmental protection

Beginning in the late 1950s and through the 1960s, Congress reacted to increasing public concern about the impact that human activity could have on the environment. A key legislative option to address this concern was the declaration of a national environmental policy. Advocates of this approach argued that without a specific policy, federal agencies were neither able nor inclined to consider the environmental impacts of their actions in fulfilling the agency's mission. The statute that ultimately addressed this issue was the National Environmental Policy Act of 1969 (NEPA, 42 U.S.C. §§ 4321–4347). Senator Henry M. Jackson proposed and helped write S 1075, the bill that eventually became the National Environmental Policy Act. The law was signed by President Nixon on January 1, 1970. NEPA was the first of several major environmental laws passed in the 1970s. It declared a national policy to protect the environment and created a Council on Environmental Quality (CEQ) in the Executive Office of the President. To implement the national policy, NEPA required that a detailed statement of environmental impacts be prepared for all major federal actions significantly affecting the environment. The "detailed statement" would ultimately be referred to as an environmental impact statement (EIS).

Ruckelshaus sworn in as first EPA Administrator.

In 1970, President Richard Nixon proposed an executive reorganization that would consolidate many of the federal government's environmental responsibilities under one agency, a new Environmental Protection Agency. That reorganization proposal was reviewed and passed by the House and Senate. For at least 10 years before NEPA was enacted, Congress debated issues that the act would ultimately address. The act was modeled on the Resources and Conservation Act of 1959, introduced by Senator James E. Murray in the 86th Congress. That bill would have established an environmental advisory counsel in the office of the President, declared a national environmental policy, and required the preparation of an annual environmental report. In the years following the introduction of Senator Murray's bill, similar bills were introduced and hearings were held to discuss the state of the environment and Congress's potential responses to perceived problems. In 1968, a joint House-Senate colloquium was convened by the chairmen of the Senate Committee on Interior and Insular Affairs (Senator Henry M. Jackson) and the House Committee on Science and Astronautics (Representative George Miller) to discuss the need for and potential means of implementing a national environmental policy. In the colloquium, some Members of Congress expressed a continuing concern over federal agency actions affecting the environment.

The EPA began regulating greenhouse gases (GHGs) from mobile and stationary sources of air pollution under the Clean Air Act (CAA) for the first time on January 2, 2011. Standards for mobile sources have been established pursuant to Section 202 of the CAA, and GHGs from stationary sources are controlled under the authority of Part C of Title I of the Act. See the page Regulation of Greenhouse Gases Under the Clean Air Act for further information.

In May 2013, Congress renamed the EPA headquarters as the William Jefferson Clinton Federal Building, after former president Bill Clinton.

Organization

Offices

- Office of the Administrator (OA)

- Office of Administration and Resources Management (OARM)

- Office of Air and Radiation (OAR)

- Office of Chemical Safety and Pollution Prevention (OCSPP)

- Office of the Chief Financial Officer (OCFO)

- Office of Enforcement and Compliance Assurance (OECA)

- Office of Environmental Information (OEI)

- Office of General Counsel (OGC)

- Office of Inspector General (OIG)

- Office of International and Tribal Affairs (OITA)

- Office of Research and Development (ORD)

- Office of Land and Emergency Management (OLEM)

- Office of Water (OW)

Regions

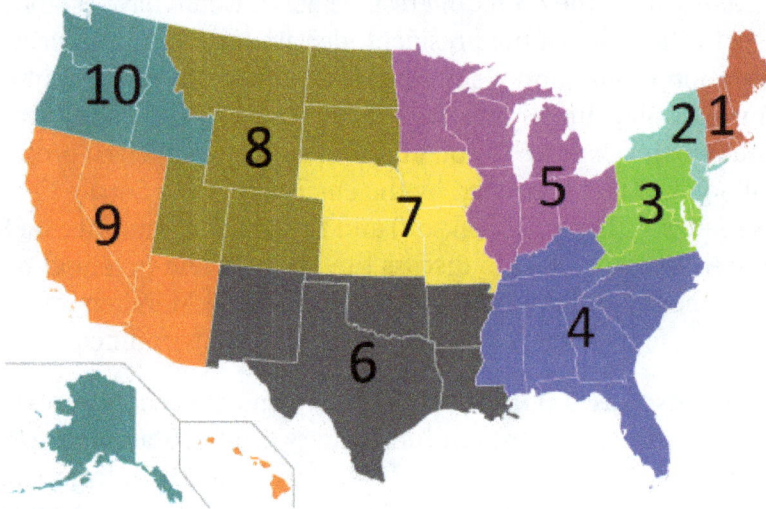

The administrative regions of the United States Environmental Protection Agency.

Each EPA regional office is responsible within its states for implementing the Agency's programs, except those programs that have been specifically delegated to states.

- Region 1: responsible within the states of Connecticut, Maine, Massachusetts, New Hampshire, Rhode Island, and Vermont (New England).

- Region 2: responsible within the states of New Jersey and New York. It is also responsible for the US territories of Puerto Rico, and the U.S. Virgin Islands.

- Region 3: responsible within the states of Delaware, Maryland, Pennsylvania, Virginia, West Virginia, and the District of Columbia.

- Region 4: responsible within the states of Alabama, Florida, Georgia, Kentucky, Mississippi, North Carolina, South Carolina, and Tennessee.

- Region 5: responsible within the states of Illinois, Indiana, Michigan, Minnesota, Ohio, and Wisconsin.

- Region 6: responsible within the states of Arkansas, Louisiana, New Mexico, Oklahoma, and Texas.

- Region 7: responsible within the states of Iowa, Kansas, Missouri, and Nebraska.

- Region 8: responsible within the states of Colorado, Montana, North Dakota, South Dakota, Utah, and Wyoming.

- Region 9: responsible within the states of Arizona, California, Hawai'i, Nevada, the territories of Guam and American Samoa, and the Navajo Nation.

- Region 10: responsible within the states of Alaska, Idaho, Oregon, and Washington.

Each regional office also implements programs on Indian Tribal lands, except those programs delegated to Tribal authorities.

Related Legislation

The legislation here is general environmental protection legislation, and may also apply to other units of the government, including the Department of the Interior and the Department of Agriculture.

Air

- 1955: Air Pollution Control Act PL 84-159

- 1963: Clean Air Act PL 88-206

- 1965: Motor Vehicle Air Pollution Control Act PL 89-272

- 1966: Clean Air Act Amendments PL 89-675

- 1967: Air Quality Act PL 90-148

- 1970: Clean Air Act Extension PL 91-604

- 1977: Clean Air Act Amendments PL 95-95

- 1990: Clean Air Act Amendments PL 101-549

Water

- 1948: Water Pollution Control Act PL 80-845

- 1965: Water Quality Act PL 89-234

- 1966: Clean Waters Restoration Act PL 89-753

- 1970: Water Quality Improvement Act PL 91-224

- 1972: Federal Water Pollution Control Amendments of 1972 PL 92-500

- 1974: Safe Drinking Water Act PL 93-523

- 1977: Clean Water Act PL 95-217

- 1987: Water Quality Act PL 100-4

- 1996: Safe Drinking Water Act Amendments of 1996

Land

- 1964: Wilderness Act PL 88-577
- 1968: Wild and Scenic Rivers Act PL 90-542
- 1970: Wilderness Act PL 91-504
- 1977: Surface Mining Control and Reclamation Act PL 95-87
- 1978: Wilderness Act PL 98-625
- 1980: Alaska National Interest Lands Conservation Act PL 96-487
- 1994: California Desert Protection Act PL 103-433
- 2010: California Desert Protection Act

Endangered Species

- 1946: Fish and Wildlife Coordination Act PL 79-732
- 1966: Endangered Species Preservation Act PL 89-669
- 1969: Endangered Species Conservation Act PL 91-135
- 1972: Marine Mammal Protection Act PL 92-522
- 1973: Endangered Species Act PL 93-205
- 1979: Endangered Species Preservation Act PL 95 335

Hazardous Waste

- 1965: Solid Waste Disposal Act PL 89-272
- 1970: Resource Recovery Act PL 91-512
- 1976: Resource Conservation and Recovery Act PL 94-580
- 1980: Comprehensive Environmental Response, Compensation, and Liability Act ("Superfund") PL 96-510
- 1984: Hazardous and Solid Wastes Amendments Act PL 98-616
- 1986: Superfund Amendments and Reauthorization Act PL 99-499
- 2002: Small Business Liability Relief and Brownfields Revitalization Act ("Brownfields Law") PL 107-118

Other

- 1947: Federal Insecticide, Fungicide, and Rodenticide Act PL 80-104
- 1969: National Environmental Policy Act PL 91-190
- 1972: Federal Environmental Pesticide Control Act PL 92-516

- 1976: Toxic Substances Control Act PL 94-469

- 1982: Nuclear Waste Repository Act PL 97-425

- 1996: Food Quality Protection Act PL 104-170

Programs

A bulldozer piles boulders in an attempt to prevent lake shore erosion, 1973 (photograph by Paul Sequeira, photojournalist and contributing photographer to the Environmental Protection Agency's DOCUMERICA project in the early 1970s)

Energy Star

In 1992 the EPA launched the Energy Star program, a voluntary program that fosters energy efficiency.

Pesticide

EPA administers the Federal Insecticide, Fungicide, and Rodenticide Act (FIFRA) (which is much older than the agency) and registers all pesticides legally sold in the United States.

Environmental Impact Statement Review

EPA is responsible for reviewing Environmental Impact Statements of other federal agencies' projects, under the National Environmental Policy Act (NEPA).

Safer Detergents Stewardship Initiative

Through the Safer Detergents Stewardship Initiative (SDSI), EPA's Design for the Environment (DfE) recognizes environmental leaders who voluntarily commit to the use of safer surfactants.

Safer surfactants are the ones that break down quickly to non-polluting compounds and help protect aquatic life in both fresh and salt water. Nonylphenol ethoxylates, commonly referred to as NPEs, are an example of a surfactant class that does not meet the definition of a safer surfactant.

The Design for the Environment, which was renamed to EPA Safer Choice in 2015, has identified safer alternative surfactants through partnerships with industry and environmental advocates. These safer alternatives are comparable in cost and are readily available. CleanGredients is a source of safer surfactants.

EPA Safer Choice

The EPA Safer Choice label, previously known as the Design for the Environment (DfE) label, helps consumers and commercial buyers identify and select products with safer chemical ingredients, without sacrificing quality or performance. When a product has the Safer Choice label, it means that every intentionally - added ingredient in the product has been evaluated by EPA scientists. Only the safest possible functional ingredients are allowed in products with the Safer Choice label.

Fuel Economy

Manufacturers selling automobiles in the United States are required to provide EPA fuel economy test results for their vehicles and the manufacturers are not allowed to provide results from alternate sources. The fuel economy is calculated using the emissions data collected during two of the vehicle's Clean Air Act certification tests by measuring the total volume of carbon captured from the exhaust during the tests.

The current testing system was originally developed in 1972 and used driving cycles designed to simulate driving during rush-hour in Los Angeles during that era. Prior to 1984 the EPA reported the exact fuel economy figures calculated from the test. In 1984, the EPA began adjusting city (aka Urban Dynamometer Driving Schedule or UDDS) results downward by 10% and highway (aka HighWay Fuel Economy Test or HWFET) results by 22% to compensate for changes in driving conditions since 1972 and to better correlate the EPA test results with real-world driving. In 1996, the EPA proposed updating the Federal Testing Procedures to add a new higher speed test (US06) and an air-conditioner on test (SC03) to further improve the correlation of fuel economy and emission estimates with real-world reports. The updated testing methodology was finalized in December, 2006 for implementation with model year 2008 vehicles and set the precedent of a 12-year review cycle for the test procedures.

In February 2005, the organization launched a program called "Your MPG" that allows drivers to add real-world fuel economy statistics into a database on the EPA's fuel economy website and compare them with others and the original EPA test results.

It is important to note that the EPA actually conducts these tests on very few vehicles. "While the public mistakenly presumes that this federal agency is hard at work conducting complicated tests on every new model of truck, van, car, and SUV, in reality, just 18 of the EPA's 17,000 employees work in the automobile-testing department in Ann Arbor, Michigan, examining 200 to 250 vehicles a year, or roughly 15 percent of new models. As to that other 85 percent, the EPA takes automakers at their word—without any testing-accepting submitted results as accurate." Two-thirds of the vehicles the EPA tests themselves are selected randomly, and the remaining third are tested for specific reasons.

Although originally created as a reference point for fossil fuelled vehicles, driving cycles have been used for estimating how many miles an electric vehicle will do on a single charge.

Air Quality

The Air Quality Modeling Group (AQMG) is in the EPA's Office of Air and Radiation (OAR) and provides leadership and direction on the full range of air quality models, air pollution dispersion models and other mathematical simulation techniques used in assessing pollution control strategies and the impacts of air pollution sources.

The AQMG serves as the focal point on air pollution modeling techniques for other EPA headquarters staff, EPA regional Offices, and State and local environmental agencies. It coordinates with the EPA's Office of Research and Development (ORD) on the development of new models and techniques, as well as wider issues of atmospheric research. Finally, the AQMG conducts modeling analyses to support the policy and regulatory decisions of the EPA's Office of Air Quality Planning and Standards (OAQPS).

The AQMG is located in Research Triangle Park, North Carolina.

Oil Pollution

The Spill Prevention, Control, and Countermeasure Rule applies to all facilities that store, handle, process, gather, transfer, store, refine, distribute, use or consume oil or oil products. Oil products includes petroleum and non-petroleum oils as well as: animal fats, oils and greases; fish and marine mammal oils; and vegetable oils, (including oils from seeds, nuts, fruits, and kernels). Mandates that a written plan is required for facilities that store more than 1,320 gallons of fuel above ground or more than 42,000 gallons below-ground, and may reasonably be expected to discharge to navigable waters(as defined in the Clean Water Act)or adjoining shorelines. Secondary Containment mandated at oil storage facilities. Oil release containment is required at oil development sites.

EPA WaterSense

WaterSense is an EPA program designed to encourage water efficiency in the United States through the use of a special label on consumer products. It was launched in June 2006. Products include high-efficiency toilets (HETs), bathroom sink faucets (and accessories), and irrigation equipment. WaterSense is a voluntary program, with EPA developing specifications for water-efficient products through a public process and product testing by independent laboratories.

Drinking Water

EPA ensures safe drinking water for the public, by setting standards for more than 160,000 public water systems nationwide. EPA oversees states, local governments and water suppliers to enforce the standards under the Safe Drinking Water Act. The program includes regulation of injection wells in order to protect underground sources of drinking water. Select readings of amounts of certain contaminants in drinking water, precipitation, and surface water, in addition to milk and air, are reported on EPA's Rad Net web site in a section entitled Envirofacts. In certain cases, readings exceeding EPA MCL levels are deleted or not included despite mandatory reporting regulations. A draft of revised EPA regulations relaxes the regulations for radiation exposure through drinking water, stating that current standards are impractical to enforce. The EPA is recommending that intervention is not necessary until drinking water is contaminated with radioactive iodine 131 at a concentration of 81,000 picocuries per liter (the limit for short term exposure set by the International Atomic Energy Agency), which is 27,000 times the current EPA limit of 3 picocuries per liter for long term exposure.

Radiation Protection

EPA has the following seven project groups to protect the public from radiation.

1. Radioactive Waste Management

2. Emergency Preparedness and Response Programs

   Protective Action Guides And Planning Guidance for Radiological Incidents

EPA developed the manual to provide guideline for local and state governments to protect public from nuclear accident.

1. EPA Cleanup and Multi-Agency Programs

2. Risk Assessment and Federal Guidance Programs

3. Naturally-Occurring Radioactive Materials Program

4. Air and Water Programs

5. Radiation Source Reduction and Management

Research Vessel

OSV *Bold* docked at Port Canaveral, FL

On March 3, 2004, the United States Navy transferred USNS *Bold*, a *Stalwart* class ocean surveillance ship, to the EPA, now known as OSV *Bold*. The ship, previously used in anti-submarine operations during the Cold War, is equipped with sidescan sonar, underwater video, water and sediment sampling instruments, used in study of ocean and coastline. One of the major missions of the *Bold* was to monitor for ecological impact sites where materials are dumped from dredging operations in U.S. ports. In 2013, the *Bold* was awarded to Seattle Central Community College (SCCC) by the General Services Administration. SCCC demonstrated in a competition that they would put it to the highest and best purpose, and acquired the ship at a cost of $5,000.

Advance Identification

Advance identification, or ADID, is a planning process used by the EPA to identify wetlands and other bodies of water and their respective suitability for the discharge of dredged and fill material. The EPA conducts the process in cooperation with the U.S. Army Corps of Engineers and local states or Native American Tribes. As of February 1993, 38 ADID projects had been completed and 33 were ongoing.

Freedom of Information Act Processing Performance

In the latest Center for Effective Government analysis of 15 federal agencies which receive the most Freedom of Information Act FOIA requests, published in 2015 (using 2012 and 2013 data, the most recent years available), the EPA earned a D by scoring 67 out of a possible 100 points, i.e. did not earn a satisfactory overall grade.

Controversies

EPA headquarters in Washington, D.C.

There has been political controversy over whether environmental regulations generally increase or decrease national employment.

The "LT2" Drinking Water Controversy

In 2005, the EPA issued the Long Term 2 Enhanced Surface Water Treatment Rule (LT2). The rule requires covering open-air reservoirs containing finished drinking water, in order to reduce the incidence of disease caused by microorganisms in drinking water.

To comply with the rule the EPA ordered that a cap be placed over the Hillview Reservoir, the open reservoir where water bound for New York City's receives its final disinfection before entering the pipelines that serve the city. A number of city and state officials complained that the project was too costly and unnecessary. The Hillview project is currently being reviewed by the EPA.

In Portland, Oregon, there was opposition to capping its reservoirs from a local citizens group and city officials, but recently the city has decided to stop fighting and bring its reservoirs into compliance with LT2.

Fiscal Mismanagement

EPA director Anne M. Gorsuch resigned under fire in 1983 during a scandal over mismanagement of a $1.6 billion program to clean up hazardous waste dumps. Gorsuch based her administration of the EPA on the New Federalism approach of downsizing federal agencies by delegating their functions and services to the individual states. She believed that the EPA was over-regulating business and that the agency was too large and not cost-effective. During her 22 months as agency head, she cut the budget of the EPA by 22%, reduced the number of cases filed against polluters, relaxed Clean Air Act regulations, and facilitated the spraying of restricted-use pesticides. She cut the total number of agency employees, and hired staff from the industries they were supposed to be regulating. Environmentalists contended that her policies were designed to placate polluters, and accused her of trying to dismantle the Agency.

In 1982 Congress charged that the EPA had mishandled the $1.6 billion toxic waste Superfund and demanded records from Gorsuch. Gorsuch refused and became the first agency director in U.S. history to be cited for contempt of Congress. The EPA turned the documents over to Congress several months later, after the White House abandoned its court claim that the documents could not be subpoened by Congress because they were covered by executive privilege. At that point, Gorsuch resigned her post, citing pressures caused by the media and the congressional investigation. Critics charged that the EPA was in a shambles at that time.

Fuel Economy

In July 2005, an EPA report showing that auto companies were using loopholes to produce less fuel-efficient cars was delayed. The report was supposed to be released the day before a controversial energy bill was passed and would have provided backup for those opposed to it, but at the last minute the EPA delayed its release.

The state of California sued the EPA for its refusal to allow California and 16 other states to raise fuel economy standards for new cars. EPA administrator Stephen L. Johnson claimed that the EPA was working on its own standards, but the move has been widely considered an attempt to shield the auto industry from environmental regulation by setting lower standards at the federal level, which would then preempt state laws. California governor Arnold Schwarzenegger, along with governors from 13 other states, stated that the EPA's actions ignored federal law, and that *existing* California standards (adopted by many states in addition to California) were almost twice as effective as the *proposed* federal standards. It was reported that Stephen Johnson ignored his own staff in making this decision.

After the federal governmtent bailed out General Motors and Chrysler in the Automotive industry crisis of 2008–2010, the 2010 Chevrolet Equinox was released with EPA fuel economy rating abnormally higher than its competitors. Independent road tests found that the vehicle did not out-perform its competitors, which had much lower fuel economy ratings. Later road tests found better, but inconclusive, results.

Palm-based biodiesel and renewable diesel failed to meet the minimum 20% greenhouse gas (GHG) emissions savings threshold requirement to qualify as renewable fuels under the US Renewable Fuel Standard 2. Palm oil plantations threaten the habitats of the endangered orang-utan and dwarf elephant.

Global Warming

In December 2007, EPA Administrator Stephen L. Johnson approved a draft of a document that declared that climate change imperiled the public welfare—a decision that would trigger the first national mandatory global-warming regulations. Associate Deputy Administrator Jason Burnett e-mailed the draft to the White House. White House aides—who had long resisted mandatory regulations as a way to address climate change—knew the gist of what Johnson's finding would be, Burnett said. They also knew that once they opened the attachment, it would become a public record, making it controversial and difficult to rescind. So they did not open it; rather, they called Johnson and asked him to take back the draft. U.S. law clearly stated that the final decision was the EPA administrator's, not President Bush's. Johnson rescinded the draft; in July 2008, he issued a new version which did not state that global warming was danger to public welfare. Burnett resigned in protest.

Libraries

In 2004, the Agency began a strategic planning exercise to develop plans for a more virtual approach to library services. The effort was curtailed in July 2005 when the Agency proposed a $2.5 million cut in its 2007 budget for libraries. Based on the proposed 2007 budget, the EPA posted a notice to the Federal Register, September 20, 2006 that EPA Headquarters Library would close its doors to walk-in patrons and visitors on October 1, 2006. The EPA also closed some of its regional libraries and reduced hours in others, using the same FY 2007 proposed budget numbers.

On October 1, 2008, the Agency re-opened regional libraries in Chicago, Dallas and Kansas City and the library at its Headquarters in Washington, DC.

In June 2011, the EPA Library Network published a strategic plan for fiscal years 2012–2014.

Mercury Emissions

In March 2005, nine states (California, New York, New Jersey, New Hampshire, Massachusetts, Maine, Connecticut, New Mexico and Vermont) sued the EPA. The EPA's inspector general had determined that the EPA's regulation of mercury emissions did not follow the Clean Air Act, and that the regulations were influenced by top political appointees. The EPA had suppressed a study it commissioned by Harvard University which contradicted its position on mercury controls. The suit alleges that the EPA's rule allowing exemption from "maximum available control technology" was illegal, and additionally charged that the EPA's system of pollution credit trading allows power plants to forego reducing mercury emissions. Several states also began to enact their own mercury emission regulations. Illinois's proposed rule would have reduced mercury emissions from power plants by an average of 90% by 2009.

9/11 Air Ratings

An August 2003 report released by EPA's Inspector General claimed that the White House put pressure on the EPA to delete cautionary information about the air quality in New York City around Ground Zero following the September 11, 2001 attacks.

An Environmental Protection Agency employee checks one of the many air sampling locations set up around the World Trade Center site.

Very Fine Airborne Particulates

Tiny particles, under 2.5 micrometres, are attributed to health and mortality concerns, so some health advocates want the EPA to regulate it. The science may be in its infancy, although many conferences have discussed the trails of this airborne matter in the air. Foreign governments such as Australia and most Member state of the European Union have addressed this issue.

The EPA first established standards in 1997, and strengthened them in 2006. As with other standards, regulation and enforcement of the $PM_{2.5}$ standards is the responsibility of the state governments, through State Implementation Plans.

Political Pressure and Scientific Integrity

In April 2008, the Union of Concerned Scientists said that more than half of the nearly 1,600 EPA staff scientists who responded online to a detailed questionnaire reported they had experienced incidents of political interference in their work. The survey included chemists, toxicologists, engineers, geologists and experts in other fields of science. About 40% of the scientists reported that the interference had been more prevalent in the last five years than in previous years. The highest number of complaints came from scientists who were involved in determining the risks of cancer by chemicals used in food and other aspects of everyday life.

EPA research has also been suppressed by career managers. Supervisors at EPA's National Center for Environmental Assessment required several paragraphs to be deleted from a peer-reviewed journal article about EPA's integrated risk information system, which led two co-authors to have their names removed from the publication, and the corresponding author, Ching-Hung Hsu, to leave EPA "because of the draconian restrictions placed on publishing". EPA subjects employees who author scientific papers to prior restraint, even if those papers are written on personal time. A $3 million mapping study on *sea level rise* was suppressed by EPA management during both the Bush and Obama Administrations, and managers changed a key interagency report to reflect the removal of the maps. EPA employees have reported difficulty in conducting and reporting the results of studies on hydraulic fracturing due to industry and governmental pressure, and are concerned about the censorship of environmental reports.

In 2015, the Government Accountability Office stated that the EPA violated federal law with covert propaganda on their social media platforms. The social media messaging that was used promoted materials supporting the Waters of the United States rule, including materials that were designed to oppose legislative efforts to limit or block the rule.

Environmental Justice

The EPA has been criticized for its lack of progress towards environmental justice. Administrator Christine Todd Whitman was criticized for her changes to President Bill Clinton's Executive Order 12898 during 2001, removing the requirements for government agencies to take the poor and minority populations into special consideration when making changes to environmental legislation, and therefore defeating the spirit of the Executive Order. In a March 2004 report, the inspector general of the agency concluded that the EPA "has not developed a clear vision or a comprehensive strategic plan, and has not established values, goals, expectations, and performance measurements" for environmental justice in its daily operations. Another report in September 2006 found the agency still had failed to review the success of its programs, policies and activities towards environmental justice. Studies have also found that poor and minority populations were underserved by the EPA's Superfund program, and that this situation was worsening.

Barriers to Enforcing Environmental Justice

Localization

Many issues of environmental justice are localized, and are therefore hard to be addressed by federal agencies such as the EPA. Without significant media attention, political interest, or 'crisis' status, local issues are less likely to be addressed at the local or federal level. With a still developing sector of environmental justice under the EPA, small, local incidents are unlikely to be solved compared to larger, well publicized incidents.

Conflicting Political Powers

The White House maintains direct control over the EPA, and its enforcements are subject to the political agenda of who is in power. Republicans and Democrats differ in their approaches to, and perceived concerns of, environmental justice. While President Bill Clinton signed the executive order 12898, the Bush administration did not develop a clear plan or establish goals for integrating environmental justice into everyday practices, which in turn affected the motivation for environmental enforcement.

Responsibilities of the EPA

The EPA is responsible for preventing and detecting environmental crimes, informing the public of environmental enforcement, and setting and monitoring standards of air pollution, water pollution, hazardous wastes and chemicals. While the EPA aids in preventing and identifying hazardous situations, it is hard to construct a specific mission statement given its wide range of responsibilities. It is impossible to address every environmental crime adequately or efficiently if there is no specific mission statement to refer to. The EPA answers to various groups, competes for resources, and confronts a wide array of harms to the environment. All of these present challenges, including a lack of resources, its self-policing policy, and a broadly defined legislation that creates too much discretion for EPA officers.

Authority of the EPA

Under different circumstances, the EPA faces many limitations to enforcing environmental justice. It does not have the authority or resources to address injustices without an increase in federal mandates requiring private industries to consider the environmental ramifications of their activities.

Gold King Mine Waste Water Spill

On August 5, 2015, while examining the level of pollutants in the Gold King Mine, EPA workers released over three million gallons of toxic waste water, including heavy metals such as lead and arsenic into Cement Creek, which flowed into the Animas River in Colorado.

Conservation International

Conservation International (CI) is an American nonprofit environmental organization headquartered in Arlington, Virginia. Its goal is to protect nature as a source of food, fresh water, livelihoods and a stable climate.

CI's work focuses on science, policy, and partnership with businesses and communities. The organization employs more than 1,000 people and works with 2,000+ partners in more than 30 countries. CI has helped establish 1,200 protected areas across 78 countries and protected more than 730 million hectares of land, marine and coastal areas (with annual Ocean Health Index).

History

Conservation International was founded in 1987 with the aim of analyzing the problems most dangerous or harmful to nature and building a foundation dedicated to solving these issues on a global scale. This model:-

- detects the problems most threatening to nature,

- prevents the industry side of the world from being detrimental to nature,

- ensures the knowledge the institution has acquired over its first twenty five years is shared with governments and, in doing so,

- establishes policies within these countries that serve as a great benefit to people and nature.

In CI's first year, the organization purchased a portion of Bolivia's foreign debt. The money was then redirected to support conservation in the Beni Biosphere Reserve. Since this first-ever debt-for-nature swap, more than $1 billion of similar deals have been made around the world.

In 1989, CI formally committed to the protection of biodiversity hotspots, ultimately identifying 34 such hotspots around the world and contributing to their protection. The model of protecting hotspots became a key way for organizations to do conservation work.

Growth and Mission Shift

In the subsequent two decades, CI expanded its work, with a stronger focus on science, corporate partnership, conservation funding, indigenous peoples, government, and marine conservation, among other things.

The organization's leadership grew to believe that CI's focus on biodiversity conservation was inadequate to protect nature and those who depended on it. CI updated its mission in 2008 to focus explicitly on the connections between human well-being and natural ecosystems.

As of FY2014, CI's expenses totaled more than US $135.3 million.

CI receives high ratings from philanthropic watchdog organizations, with an A rating from Charity Watch. Charity Navigator awarded CI a score of 92.28 out of 100 for accountability and transparency.

Approach to Conservation

The foundation of CI's work is "science, partnership and field demonstration." The organization has scientists, policy workers and other conservationists on the ground in more than 30 countries. It also relies heavily on thousands of local partners.

CI works with governments, universities, NGOs and the private sector with the aim of replicating these successes on a larger scale. By showing how conservation can work at all scales, CI aims to make the protection of nature a key consideration in economic development decisions around the world. CI supported 23 Pacific Island nations and territories in the formation of the Pacific Oceanscape, a management plan for the conservation of nearly 24 million square miles of sea from Hawaii to New Zealand. In addition to the sustainable management of ocean resources, the agreement includes the world's largest marine protected areas and sanctuaries for whales, dolphins, turtles and sharks.

The organization has been active in United Nations discussions on issues such as climate change and biodiversity, and its scientists present at international conferences and workshops. Its United States policy work currently highlights "a direct connection between international conservation and America's economic and national security interests."

A few years after its founding, CI began working with McDonald's to implement sustainable agriculture and conservation projects in Central America. The organization expanded its commitment to working with the business sector in 2000, when it created the Center for Environmental Leadership in Business with support from the Ford Motor Company.

Criticism

CI has been criticized for links to companies with a poor environmental record such as BP, Cargill, Chevron, Monsanto and Shell and for allegedly offering greenwashing services.

A 2008 article in The Nation claimed that the organization had attracted $6 million for marine conservation in Papua New Guinea, but that the funds were used for "little more than plush offices and first class travel."

In 2011, Conservation International was targeted by a group of reporters from Don't Panic TV who posed as an American arms company and asked if the charity could "raise [their] green profile." Options outlined by the representative of Conservation International (CI) included assisting with the arms company's green PR efforts, membership of a business forum in return for a fee, and sponsorship packages where the arms company could potentially invest money in return for being associated with conservation activities. Conservation International agreed to help the arms company find an "endangered species mascot." Film footage shows the Conservation International employee suggesting a vulture North African birds of prey as a possible endangered species mascot for the arms company. CI contends that these recordings were heavily edited to remove elements that would have cast CI in a more favorable light, while using other parts of the video out of context to paint an inaccurate and incomplete picture of CI's work with the private sector.

In March 2012 The Phnom Penh Post reported that forest rangers appointed by Conservation International had accepted bribes, and that a CI employee who brought this to the attention of CI was fired.

In May and June 2013, Survival International reported that an indigenous Bushman tribe in Botswana was threatened with eviction from their ancestral land in order to create a wildlife corridor known as the Western Kgalagadi Conservation Corridor. A Botswana government representative denied this. A May press release from CI said, "Contrary to recent reports, Conservation International (CI) has not been involved in the implementation of conservation corridors in Botswana since 2011", and asserted that CI had always supported the San Bushmen and their rights.

Leadership

- Chairman and CEO: Peter Seligmann

- Chief Operating Officer: Jennifer Morris

- Chairman of the Executive Committee: Rob Walton, chairman of the board,

- Vice Chairs: Harrison Ford, actor and André Esteves, CEO, Banco BTG Pactual S/A, São Paulo, Brazil.

- Executive Vice Chair, Dr. Russell Mittermeier

References

- *Aboul-Enein, H. Yousuf; Zuhur, Sherifa (2004), Islamic Rulings on Warfare, Strategic Studies Institute, US Army War College, Diane Publishing Co., Darby PA, p. 22, ISBN 9781584871774*

- *Thomas R. DeGregori (2002). Bountiful Harvest: Technology, Food Safety, and the Environment. Cato Institute. p. 153. ISBN 1-930865-31-7.*

- *Craig Kridel (2010-02-16). Encyclopedia of Curriculum Studies. Sage Publications, Inc. p. 341. ISBN 978-1-4129-5883-7. Retrieved 2010-04-16.*

- Huesemann, Michael H., and Joyce A. Huesemann (2011). *Technofix: Why Technology Won't Save Us or the Environment*, New Society Publishers, Gabriola Island, British Columbia, Canada, ISBN 0865717044, 464 pp.

- Dickens, P. 2004, *Society & Nature: Changing Our Environment, Changing Ourselves*, Cambridge, UK, Polity, ISBN 0-7456-2796-X.

- Hajer, M.A., 1995, *The Politics of Environmental Discourse: Ecological Modernization and the Policy Process*, Oxford, UK, Oxford University Press, ISBN 0-19-827969-8.

- Mol, A.P.J., and Sonnenfeld, D.A., (eds.) 2000, *Ecological Modernisation around the World: Perspectives and Critical Debates*, London and Portland, OR, Frank Cass/ Routledge, ISBN 978-0-7146-8113-9.

- Mol, A.P.J., Sonnenfeld, D.A., and Spaargaren, G., (eds.) 2009, *The Ecological Modernisation Reader: Environmental Reform in Theory and Practice*, London and New York, Routledge, ISBN 978-0-415-45370-7 hardback, ISBN 978-0-415-45371-4 paperback.

- Redclift, M. R., and Woodgate, G. (eds.) 1997, *The International Handbook of Environmental Sociology*, Cheltenham, UK, Edward Elgar, ISBN 1-85898-405-X.

- Redclift, M. R., and Woodgate, G., (eds.) 2005, *New Developments in Environmental Sociology*, Cheltenham, Edward Elgar, ISBN 1-84376-115-7.

- Young, S.C., 2000, *The Emergence of Ecological Modernisation: Integrating the Environment and the Economy?*, London, Routledge, ISBN 0-415-14173-7.

- *Mautner, Michael N. (2000). Seeding the Universe with Life: Securing Our Cosmological Future (PDF). Washington D. C.: Legacy Books. ISBN 0-476-00330-X.*

- *Lindholt, Lone (2005). Human Rights in Development Yearbook 2003: Human Rights and Local/living Law. Martinus Nijhoff Publishers. ISBN 90-04-13876-5.*

- *Gray, Andrew (2003). Indigenous Rights and Development: Self-Determination in an Amazonian Community. Berghahn Books. ISBN 1-57181-837-5.*

- *Keal, Paul (2003). European Conquest and the Rights of Indigenous Peoples: The Moral Backwardness of International Society. Cambridge University Press.ISBN 0-521-82471-0.*

Permissions

All chapters in this book are published with permission under the Creative Commons Attribution Share Alike License or equivalent. Every chapter published in this book has been scrutinized by our experts. Their significance has been extensively debated. The topics covered herein carry significant information for a comprehensive understanding. They may even be implemented as practical applications or may be referred to as a beginning point for further studies.

We would like to thank the editorial team for lending their expertise to make the book truly unique. They have played a crucial role in the development of this book. Without their invaluable contributions this book wouldn't have been possible. They have made vital efforts to compile up to date information on the varied aspects of this subject to make this book a valuable addition to the collection of many professionals and students.

This book was conceptualized with the vision of imparting up-to-date and integrated information in this field. To ensure the same, a matchless editorial board was set up. Every individual on the board went through rigorous rounds of assessment to prove their worth. After which they invested a large part of their time researching and compiling the most relevant data for our readers.

The editorial board has been involved in producing this book since its inception. They have spent rigorous hours researching and exploring the diverse topics which have resulted in the successful publishing of this book. They have passed on their knowledge of decades through this book. To expedite this challenging task, the publisher supported the team at every step. A small team of assistant editors was also appointed to further simplify the editing procedure and attain best results for the readers.

Apart from the editorial board, the designing team has also invested a significant amount of their time in understanding the subject and creating the most relevant covers. They scrutinized every image to scout for the most suitable representation of the subject and create an appropriate cover for the book.

The publishing team has been an ardent support to the editorial, designing and production team. Their endless efforts to recruit the best for this project, has resulted in the accomplishment of this book. They are a veteran in the field of academics and their pool of knowledge is as vast as their experience in printing. Their expertise and guidance has proved useful at every step. Their uncompromising quality standards have made this book an exceptional effort. Their encouragement from time to time has been an inspiration for everyone.

The publisher and the editorial board hope that this book will prove to be a valuable piece of knowledge for students, practitioners and scholars across the globe.

Index

www.ingramcontent.com/pod-product-compliance
Lightning Source LLC
Chambersburg PA
CBHW061316190326
41458CB00011B/3825